William Chauvenet

New Method of Correcting Lunar Distances

And Improved Method of Finding the Error and Rate of a Chronometer by

Equal Altitudes

William Chauvenet

New Method of Correcting Lunar Distances
And Improved Method of Finding the Error and Rate of a Chronometer by Equal Altitudes

ISBN/EAN: 9783337399184

Printed in Europe, USA, Canada, Australia, Japan

Cover: Foto ©berggeist007 / pixelio.de

More available books at **www.hansebooks.com**

OF

CORRECTING LUNAR DISTANCES,

AND

IMPROVED METHOD

OF FINDING THE

ERROR AND RATE OF A CHRONOMETER

BY

EQUAL ALTITUDES.

By WILLIAM CHAUVENET, LL. D.,

CHANCELLOR OF WASHINGTON UNIVERSITY, ST. LOUIS, LATE PROFESSOR OF MATHEMATICS AND ASTRONOMY
IN THE U. S. NAVAL ACADEMY, ETC.

EXTRACTED FROM THE APPENDIX TO THE AMERICAN EPHEMERIS FOR 1857, BY AUTHORITY OF THE
BUREAU OF NAVIGATION, NAVY DEPARTMENT, WASHINGTON.

NEW YORK:
D. VAN NOSTRAND, 192 BROADWAY.
LONDON: TRÜBNER & CO.

1868.

NEW METHOD

OF

CORRECTING LUNAR DISTANCES.

FROM THE APPENDIX TO THE AMERICAN EPHEMERIS AND NAUTICAL
ALMANAC FOR 1855.

DIRECTIONS

THE TABLES FOR CORRECTING LUNAR DISTANCES.

THE object of these Tables is to give the *true* correction of a lunar distance in all cases when, with the apparent distance of the moon from the sun, a planet, or star, the apparent altitudes of the two objects have also been obtained by observation. They enable us readily to take into account, — 1st, the parallax of the moon in the latitude of the observer, allowing for the spheroidal figure of the earth; 2d, the parallax of the sun or a planet; 3d, the true atmospheric refraction, allowing for the actual state of the air as shown by the barometer and thermometer; and 4th, that effect of refraction which gives the apparent discs of the moon and sun an oval or elliptical figure.

The longitude deduced from a lunar observation, when no attention is paid to the spheroidal figure of the earth, to the barometer and thermometer, and the elliptical figure of the discs, may in certain cases be in error *a whole degree.* It is true these extreme cases are rare in practice; but cases are common in which from such neglect the error in the longitude is 10′, 15′, or 20′. Since lunars are now chiefly valuable as checks upon the chronometer, it is absolutely necessary to get rid of such errors, and to leave no other inaccuracy in the result than that which unavoidably follows from the observations. This is accomplished by means of these Tables, with an amount of labor very little greater than that which is required by the inaccurate methods in common use.

THE OBSERVATION.

The record of a *complete* observation embraces, —

1. The latitude and approximate longitude of the place of observation.

2. The approximate local time.

3. The time of observation as shown by a chronometer, and the error of the chronometer, or its difference from mean Greenwich time.

4. The apparent distance of the moon's bright limb from a star or planet, or from the nearest limb of the sun.

5. The apparent altitude of the moon's upper or lower limb above the sea horizon.

6. The apparent altitude of the star, planet, or lower limb of the sun above the sea horizon.

5

7. The height of the barometer and thermometer.

8. The height of the eye above the level of the sea.

9. The index correction of the sextant, if a sextant is used.

The index correction of the sextant may be supposed to be previously determined, but, since even in the best instruments it is not constant, its determination should be considered a necessary part of the observation; and when the greatest precision is sought, it should be found both before and after the measurement of the distance, and its mean value taken.

The error of the chronometer above alluded to is that which is obtained by applying the daily rate (multiplied by the proper number of days) to the error found before leaving port. The agreement or disagreement of the error thus found with that found by the lunar observation will be the test of the good or bad going of the chronometer.

PREPARATION OF THE DATA.

Greenwich Date. — Correct the chronometer time for its error from Greenwich time and deduce the Greenwich Date, i. e. the Greenwich day and hour (mean time), reckoning the hours in succession from 0 to 24, beginning at noon.

Nautical Almanac. — With the Greenwich Date enter the Almanac and take out the moon's semidiameter and horizontal parallax; and if the sun is observed, its semidiameter and horizontal parallax; * but if a planet is observed, its horizontal parallax only.

Apparent Altitude of the Moon. — To the altitude given by the sextant apply the index correction of the instrument and subtract the dip of the horizon, Table I. If the lower limb is observed, add the semidiameter augmented by Table II.; if the upper limb is observed, subtract the augmented semidiameter. The result is the apparent altitude of the moon's centre, denoted " $\mathbb{C}$'s *App. Alt.*"

Apparent Altitude of the Sun, Planet, or Star. — To the observed altitude apply the index correction of the sextant, and subtract the dip, Table I.; and if the sun is used, add its semidiameter when the lower limb is observed, or subtract it when the upper limb is observed. The result is the apparent altitude required, denoted by " $\odot$'s or $*$'s *App. Alt.*"

Apparent Distance. — 1st. When the sun is used, to the observed distance (corrected for index error when necessary) add the moon's augmented semidiameter and the sun's semidiameter. 2d. When a planet or star is used, add the moon's augmented semidiameter if its nearest limb is observed, but subtract it if its farthest limb is observed. The result is " *App. Dist.*"

Moon's Reduced Parallax and Refraction. — Enter Table III. with the latitude of the place of observation and the moon's horizontal parallax, and take out the correction, which add to the horizontal parallax. Call the result the moon's reduced parallax, or " $\mathbb{C}$'s *Red. P.*"

Enter Table IV. with the moon's apparent altitude, and take out the mean reduced

---

* The sun's horizontal parallax may be assumed as $8''.5$.

refraction, and apply to this mean refraction the corrections given in Tables IV. A. and IV. B., adding or subtracting these corrections according to the directions in the Tables. The result is the moon's reduced refraction or " ☾'s *Red. R.*"

Subtract the " ☾'s *Red. R.*" from the " ☾'s *Red. P.*" and mark the result as " ☾'s *Red. P. and R.*"

*Reduced Parallax and Refraction of Sun, Planet, or Star.** — With the apparent altitude of the sun, planet, or star, take from Table IV. the mean reduced refraction, which correct by Tables IV. A. and IV. B. If the sun is observed, subtract its horizontal parallax (which may be always taken at 8″.5) from its reduced refraction, and mark the result as " ☉'s *Red. P. & R.*" If a planet is observed subtract its horizontal parallax, and mark the result as " ✶'s *Red. P. & R.*" If a star is observed its reduced refraction is at once the required " ✶'s *Red. P. & R.*"

COMPUTATION OF THE TRUE DISTANCE.

Take from Table V. the four logarithms *A, B, C, D*,† and place these logs. each at the head of a column, marking the columns *A, B, C,* and *D*, respectively ; then put the

log. of ☾'s *Red. P. & R.*	(Table IX.)	in Columns *A* and *B*
log. of ☉'s *Red. P. & R.*	"	" *C* " *D*
log. sine ☾'s *App. Alt.*	(Bowd. Table XXVII.)	" *A* " *D*
log. sine ☉'s *App. Alt.*	"	" *B* " *C*
log. cotangent *App. Dist.*	"	" *A* " *C*
log. cosecant *App. Dist.*	"	" *B* " *D.*

The sum of the four logs. in Col. *A* is the log. (Table IX.) of the *First Part of* ☾'s *Correction*, which is to be marked + when the App. Dist. is less than 90°, but — when the App. Dist. is greater than 90°.

The sum of the four logs. in Col. *B* is the log. (Table IX.) of the *Second Part of* ☾'s *Correction*, which is always to be marked —.

The sum of the four logs. in Col. *C* is the log. (Table IX.) of the *First Part of the* ☉'s *or* ✶'s *Correction*, which is to be marked — when the App. Dist. is less than 90°, but + when the App. Dist. is greater than 90°.

The sum of the four logs. in Col. *D* is the log. (Table IX.) of the *Second Part of the* ☉'s *or* ✶'s *Correction*, which is always to be marked +.

Combine the first and second parts of the ☾'s correction according to the signs (+ or —) prefixed ; that is, take their *sum* if they have the *same* sign, but their difference if they have *different* signs, and prefix the sign of the greater to the result, which call " ☾'s *whole Correction.*"

In the same manner form the ☉'s *or* ✶'s *whole Correction.*

First Correction of Distance. — Combine the ☾'s *whole corr.* and the ☉'s *or* ✶'s

---

* The parallax of a star being zero, its "reduced parallax and refraction" become, of course, merely its "reduced refraction"; but as no mistake can arise from marking it as "✶'s *P. & R.*," this designation has been retained in order to give simplicity and uniformity at once to the rules and the tables.

† No interpolation is required in taking out these logarithms.

whole corr. according to their signs; the result is the *First Correction of Distance*, which is to be added to or subtracted from the apparent distance, according as its sign is + or —.

Second Correction of Distance. — Enter Table VI. with the Apparent Distance and the First Correction of Distance, and take out the *Second Correction of Distance*, which is to be applied to the distance according to the directions in the side columns of the Table.

Correction for the Elliptical Figure of the Moon's Disc, or Contraction of the Moon's Semidiameter (Table VII.). — Enter Table VII. A. with the ☾'s App. Alt. and ☾'s Red. P. & R., and take out the number. With this number and the ☾'s whole correction enter Table VII. B. and take the required *contraction*, which is to be *added* to the App. Dist. when the *farthest* limb is observed, but *subtracted* when the *nearest* limb is observed.

Correction for the Elliptical Figure of the Sun's Disc, or Contraction of the Sun's Semidiameter (Table VIII.). — Enter Table VIII. A. with the ☉'s App. Alt. and ☉'s Red. P. & R. and take out the number. With this number and the ☉'s whole corr. enter Table VIII. B. and take out the required *contraction*, which is always to be *subtracted* from the distance (the *nearest* limb of the sun being always observed).

Correction for Compression or for the Spheroidal Figure of the Earth. — Take from the Nautical Almanac for the Greenwich Date the value of log. *N*, given with the lunar distance observed.* To log. *N* add the log. sine of the latitude of the place of observation; the sum is the log. (Table IX.) of the required *correction for compression.* In North latitude *add* this correction to the distance if log. *N* in the Nautical Almanac is marked +, or *subtract* it if log. *N* is marked — ; in South latitude subtract the correction when log. *N* is +, and add it when log. *N* is —.

All these corrections being applied to the Apparent Distance, the result is the *True Distance.*

To find the Error of the Chronometer.

Find in the Nautical Almanac the two distances between which the true distance falls. Take out the first of these together with the Prop. Log. following it and the hours of Greenwich time over it. Find the difference between the distance taken from the Almanac and the true distance, and to the log. of this difference (Table IX.) *add* the Prop. Log. from the Almanac; the sum is the log. (Table IX.) of an interval

---

* The values of log. *N* are no longer given in the Nautical Almanac. For distances of the moon from the sun for the year 1855, they are given in Table XI. of this volume. For other distances, at any time, log. *N* may be found from Table XII. by the following rule :

From the Nautical Almanac take the moon's and star's (or planet's) declinations to the nearest whole degree. With the moon's declination and apparent distance take from Table XII. A. the *first part of N*, and mark it with the sign in the table if the declination is *North*; but if the declination is *South*, change the sign from + to — or from — to +. With the star's (or planet's) declination and apparent distance, take from Table XII. B. the *second part of N*, and mark it + when the declination is *North*, and — when *South*. Take the *sum*, or *difference*, of the two parts, according as their signs are the *same* or *different*, and to the resulting number prefix the sign of the greater. The logarithm of this number of seconds, taken in Table IX., with its sign prefixed, is the required log. *N*, to be used as directed in the text.

of time to be added to the hours of Greenwich time taken from the Almanac. **The** result is the *approximate* Greenwich time.

To correct this Greenwich time, take the difference between the two Prop. Logs. in the Almanac which stand against the two distances between which the true distance falls. With this difference and the interval of time just found, enter Table X. of this Appendix, and take out the seconds, which are to be *added* to the approximate Greenwich time when the Prop. Logs. are *decreasing*, but *subtracted* when the Prop. Logs. are *increasing*. The result is the *true Greenwich time.*

The difference between the true Greenwich time and the time shown by the chronometer is the error of the chronometer as determined by the lunar observation.

DEGREE OF DEPENDENCE.

If the error thus determined agrees with that deduced by means of the rate and original error, the chronometer has run well, and its rate is confirmed; if otherwise, more or less doubt is thrown upon the chronometer, according to the degree of accuracy of the lunar observation itself. An error of $10''$ in the measurement of the distance produces about 20^s error in the Greenwich time; and since, even with the best observers, a single set of distances is subject to a possible error of $10''$, it may be well to consider the chronometer as still to be trusted so long as it does not differ from the lunar by more than 20^s. Since, however, so much depends upon skill in measuring the distance, the observer can only form a correct judgment of the degree of dependence to be placed upon his own observations by repeated trials and a careful comparison of his several results.

CORRECTION OF

EXAMPLE 1.

In Lat. 35° 30′ N., Long. 30° W., by account, on September 7th, 1855, about 6ʰ A. M., the Greenwich chronometer showing 8ʰ· 29ᵐ· 57ˢ·.5 and supposed to be *fast* 21ᵐ· 1ˢ·.5 ; the observed distance of ☉'s and ☾'s nearest limbs is 43° 52′ 10″ ; observed alt. ☾ 49° 32′ 50″ ; observed alt. ☉ 5° 27′ 10″ ; barometer 29.1 inches, thermometer 75° ; height of the eye above the sea 20 feet ; index correction of the sextant 0. What is the error of the chronometer on Greenwich time according to these observations ?

Preparation of the Data.

	h. m. s.				
Chronometer	8 29 57.5	☾'s semid. N. A.	14 50.0	☉'s Par. N. A.	54 19.4
(Fast)	— 21 1.5	Aug. Tab. II.	+11.2	Aug. Tab. III.	+3.6
Greenw. date Sept. 6	20 8 56.0	☾'s aug. Semid.	15 1.2	☾'s Red. P.	54 23.0

	° ′ ″		° ′ ″		° ′ ″
Obs'd Alt. ☾	49 32 50	Obs'd alt. ☉	5 27 10	Obs'd distance ☉| |☾	43 52 10
Dip Tab. I.	— 4 23	Dip	— 4 23	☾'s aug. semid.	+ 15 1
☾'s aug. semid.	+15 1	☉'s semid	+15 55	☉'s semid.	+15 55
☾'s *App. Alt.*	49 43 28	☉'s *App. Alt.*	5 38 42	*App. Dist.*	44 23 6

	′ ″		′ ″
☾'s Red. R. Tab. IV.	1 16	☉'s Red. R. Tab. IV.	8 57
Barom. Tab. IV. A.	—3	Barom. Tab. IV. A.	—16
Therm. Tab. IV. B.	—4	Therm. Tab. IV. B.	—28
☾'s *Red. R.*	1 9	☉'s *Red. R.*	8 13
☾'s *Red. P.*	54 23	☉'s *Par.*	8
☾'s *Red. P. & R.*	53 14	☉'s *Red. P. & R.*	8 5

Computation of the True Distance.

A.

log. A. Tab. V.	0.0021
log. ☾'s Red. P. & R. Tab. IX.	3.5043
log. sin. ☾'s App. Alt.	9.8825
log. cot. App. Dist.	0.0093
{ log. Tab. IX.	3.3982
{ 1st Part ☾'s corr.	+41′ 42″

C.

log. C. Tab. V.	9.9949
log. ☉'s Red. P. & R. Tab. IX.	2.6857
log. sin. ☉'s App. Alt.	8.9929
log. cot. App. Dist.	0.0093
{ log. Tab. IX.	1.6828
{ 1st Part ☉'s corr.	—0′ 48″

B.

log. B. Tab. V.	9.9951
log. ☾'s Red. P. & R.	3.5043
log. sin. ☉'s App. Alt.	8.9929
log. cosec. App. Dist.	0.1552
{ log. Tab. IX.	2.6475
{ 2d Part ☾'s corr.	— 7′ 24″
☾'s *whole corr.*	+34′ 18″

D.

log. D. Tab. V.	9.9992
log. ☉'s Red. P & R.	2.6857
log. sin. ☾'s App. Alt.	9.8825
log. cosec. App. Dist.	0.1552
{ log. Tab. IX.	2.7226
{ 2d Part ☉'s corr.	+8′ 48″
☉'s *whole corr.*	+8′ 0″

	° ′ ″
App. Dist.	44 23 6
1st corr.	+42 18
2d corr. Tab. VI	— 16
Contraction of ☾'s *Semid.* Tab. VII.	0
Contraction of ☉'s *Semid.* Tab. VIII.	— 20
Corr. for compression	— 3
True Distance	45 4 45

log. N.* Tab. XI.	—0.764
log. sin. Lat. 35° 30′ N.	+9.764
log. Tab. IX.	—0.528

. . . .

---

* This log. may also be found from Table XII. by the rule in the note on p. 8.

To find the Error of the Chronometer.

True distance	45° 4′ 45″			
Distance N. A. at XVIII.ʰ·	46 3 17	P. L.	0.3433	Diff. P. logs. +5
Difference	58 32	log. Tab. IX. 3.5456		
Approximate interval	2ʰ· 9ᵐ· 3ˢ·	log. Tab. IX. 3.8889		
Add	18			
Approx. Gr. time	20 9 3			
Corr. Tab. X.	−2			
True Gr. time	20 9 1			
Gr. time by Chronom.	20 8 56			
Chronom. and lunar differ only	5			

and therefore the chronometer may be considered as going well.

This example, worked by Bowditch's First Method, gives the true distance 45° 5′ 44″, differing from the above 59″, in consequence of the omission by Bowditch of the small corrections. This difference would produce an error of 2ᵐ· 10ˢ· in the Greenwich time, and consequently the longitude in this case deduced by Bowditch's method would be in error 32′.5; that is, *more than half a degree.*

EXAMPLE II.

In Lat. 55° 20′ S., Long. 120° 25′ W., by account, on August 29th, 1855, about 9ʰ· 40ᵐ· P. M., the Greenwich chronometer showing 5ʰ· 35ᵐ· 46ˢ·.2 and from previous rate supposed *slow* 5ᵐ· 12ˢ·; the following distance and altitudes are found, being the mean of six observations corrected for index errors. Observed distance of Fomalhaut and ☽'s farthest limb 46° 30′ 23″; observed alt. ☽ 6° 26′ 10″; observed alt. Fomalhaut 52° 34′ 40″; barometer 31 inches; thermometer 20°; height of the eye above the sea 18 feet. What is the error of the chronometer according to these observations?

Preparation of the Data.

	h. m. s.		″		′ ″
Chronometer	5 35 46.2	☽'s semid. N. A.	16 26.3	☽'s Par. N. A.	60 11.8
(Slow)	+ 5 12	Aug. Tab. II.	+2.0	Aug. Tab. III.	+8.3
Greenw. date Aug. 29ᵈ·	17 40 58.2	☽'s aug. Semid.	16 28.3	☽'s Red. P.	60 20.1

	° ′ ″		° ′ ″		° ′ ″
Obs'd Alt. ☽	6 26 10	Obs'd alt. ✳	52 34 40	Obs'd distance ✳ ☽	46 30 23
Dip Tab. I.	− 4 9	Dip	− 4 9	☽'s aug. semid.	− 16 28
☽'s aug. semid.	+16 28	✳'s App. Alt.	52 30 31	App. Dist.	46 13 55
☽'s App. Alt.	6 38 29				

	′ ″		′ ″
☽'s Red. R. Tab. IV.	7 48	✳'s Red. R. Tab. IV.	1 13
Barom. Tab. IV. A.	+16	Barom. Tab. IV. A.	+ 2
Therm. Tab. IV. B.	+32	Therm. Tab. IV. B.	+ 5
☽'s Red. R.	8 36	✳'s Red. R.	1 20
☽'s Red. P.	60 20	✳'s Par.	0
☽'s Red. P. & R.	51 44	✳'s Red. P. & R.	1 20

Computation of the True Distance.

A.	
log. A. Tab. V.	0.0274
log. ☾'s Red. P.& R. Tab.IX.	3.4919
log. sin. ☾'s App. Alt.	9.0632
log. cot. App. Dist.	9.9813
⎰ log. Tab. IX.	2.5638
⎱ 1st *Part* ☾'s *corr.*	+6' 6''

C.	
log. C. Tab. V.	9.9999
log.✴'s Red.P.& R. Tab.IX.	1.9031
log. sin. ✴'s App. Alt.	9.8995
log. cot. App. Dist.	9.9813
⎰ log. Tab. IX.	1.7838
⎱ 1st *Part* ✴'s *corr.*	—1' 1''

B.	
log. B. Tab. V.	0.0001
log. ☾'s Red. P. & R.	3.4919
log. sin. ✴'s App. Alt.	9.8995
log. cosec. App. Dist.	0.1414
⎰ log. Tab. IX.	3.5329
⎱ 2d *Part* ☾'s *corr.*	—56' 51''
☾'s *whole corr.*	—50' 45''

D.	
log. D. Tab. V.	0.0267
log. ✴'s Red. P. & R.	1.9031
log. sin. ☾'s App. Alt.	9.0632
log. cosec. App. Dist.	0.1414
⎰ log. Tab. IX.	1.1344
⎱ 2d *Part* ✴'s *corr.*	+0' 14''
✴'s *whole corr.*	—0' 47''

log. N.*	—1.230
log. sin. Lat. 55° S.	—9.913
log. Tab. IX.	+1.143

. . . .

	2 '	
App. Dist.	46 13 55	
1st *corr.*	—51 32	
2d *corr.* Tab. VI.	— 22	
Contraction of ☾'s ⎱	+ 17	
Semid. Tab. VII. ⎰		
Corr. for compression	+ 14	
True Distance	45 22 32	

To find the Error of the Chronometer.

True Distance	45° 22' 32''			
Dist. N. A. at X V.ʰ·	43 51 59	P. L.	0.2527	Diff. P. Logs. —6
Difference	1 30 33	log. Tab. IX.	3.7350	
Approx. interval	2ʰ· 42ᵐ· 1ˢ·	log. Tab. IX.	3.9877	
Add	15			
Approx. Gr. time	17 42 1			
Corr. Tab. X.	+1			
True Gr. time	17 42 2			
Gr. time by chronom.	17 40 58			
Chron. and lunar differ	1 4			

and, the distances having been observed with care, the chronometer has probably changed its rate. A second observation confirming this, we must, from repeated lunars, determine a new rate, which may be used until an opportunity occurs of rating at a fixed place whose longitude is tolerably well known.

This example, worked by Bowditch's Second Method, gives the true distance 45° 21' 31'', which is in error 1' 1'', and would produce in the longitude deduced from it in this case an error of about 28'.

* This value of log. N. is formed from Tab. XII. by the rule in the note on page 8 of this Appendix: thus

☾'s dec. 4° N., App. Dist. 46° in Tab. XII. A. gives 1st Part of N — 1''

✴'s dec. 30° S., App. Dist. 46° in Tab. XII. B. gives 2d Part of N —16

$$\overline{\qquad\qquad} \text{N } \overline{—17}$$

the log. of which is in Tab. IX. 1.230.

EXPLANATION OF THE TABLES.*

TABLE I. — *Dip of the Sea Horizon*, computed by Delambre's formula (*Astrono·mie*, Vol. III., Chap. XXXVI.), which, when feet are substituted for metres, is

$$D = 58''.8 \sqrt{F},$$

where $F =$ height of the eye above the level of the sea, in feet,
   $D =$ depression of the sea horizon, in seconds.

TABLE II. — *Augmentation of the Moon's Semidiameter*, computed by the formula (Francœur, *Astron. Pratique*, p. 58)

$$x = c \, s^2 \sin h + \tfrac{1}{2} c^2 s^3 \sin^2 h + \tfrac{1}{2} c^2 s^3,$$

where $h =$ moon's apparent altitude,
   $s =$ moon's horizontal semidiameter,
   $x =$ augmentation of semidiameter for the altitude h,
   $\log c = 5.25021.$

TABLE III. — *Augmentation of the Moon's Horizontal Parallax*, or correction to reduce the moon's equatorial horizontal parallax to that point of the earth's axis which lies in the vertical of the observer in any given latitude, computed by the formulas

$$\Delta \pi = \pi \, (b - 1), \quad b = \frac{1}{\sqrt{(1 - e^2 \sin^2 \phi)}},$$

where $\pi =$ equatorial horizontal parallax,
   $\phi =$ latitude,
   $e =$ eccentricity of the meridian, $\log e^2 = 7.81602,$
   $\Delta \pi =$ augmentation of the horizontal parallax for the latitude ϕ.

TABLE IV. — *Mean Reduced Refraction for Lunars*, computed by the formula

$$r' = \frac{r}{\cos h} = \frac{k}{\sin h},$$

---

* Tables I., II., and IX., which are not peculiar to my method, are inserted to save a reference to other works for them; so that the computer requires, in connection with this volume, only a table of logarithmic sines and tangents. These three tables have, however, been recomputed, as stated in the "Explanation."

13

where h = the apparent altitude,

r = mean refraction, barometer 30 inches, Fahrenheit's thermometer 50°,

r' = mean reduced refraction for lunars.

The refractions employed are Bessel's, and were taken from his table (*Astronomische Untersuchungen*, Vol. I. p. 200), which gives directly $K = r \tan h$.

Tables IV. A. and IV. B. — *Corrections of the Mean Refraction for the Height of the Barometer and Thermometer*, deduced also from Bessel's table above cited. These tables serve for correcting either r or r', according as the one or the other is taken as the argument.

Table V. — *Logs. of A, B, C, and D, for computing the First Correction of the Lunar Distance*, computed by the formulas

$$A = K'^2 \frac{\sin (h + \frac{1}{2}\,\Delta\,h)}{\sin h}, \qquad C = \frac{\sin (H - \Delta\,H)}{\sin H},$$

$$B = K' \frac{\sin (2\,H - \Delta\,H)}{\sin 2\,H}, \qquad D = \frac{\sin (2\,h + \Delta\,h)}{\sin 2\,h},$$

where h = moon's apparent altitude,

H = sun's, planet's, or star's apparent altitude (denoted in the tables by $\odot$'s or $*$'s App. Alt.),

$\Delta\,h$ = difference of $\mathbb{C}$'s apparent and true altitudes,

$\Delta\,H$ = difference of $\odot$'s or $*$'s apparent and true altitudes,

$\log K' = .000126,$

and $\Delta\,h$, $\Delta\,H$ were computed from the arguments "apparent altitude" and "reduced parallax and refraction" by the formulas

$$\Delta\,h = (p - r')\,K' \cos h, \qquad \Delta\,H = (R' - P) \cos H,$$

where $p - r'$ = $\mathbb{C}$'s reduced parallax and refraction,

$R' - P$ = $\odot$'s or $*$'s reduced parallax and refraction,

p = $\mathbb{C}$'s horizontal parallax $+ \Delta\,\pi$ (Table III.),

P = $\odot$'s or $*$'s horizontal parallax (for a star $P = 0$),

r' = $\mathbb{C}$'s reduced refraction (Table IV.),

R' = $\odot$'s or $*$'s reduced refraction (Table IV.).

When h and H become 90°, the values of B and D assume the indeterminate forms

$$B = \frac{0}{0}, \qquad D = \frac{0}{0},$$

and therefore, for computing their logarithms near the end of the table, the formulas were transformed as follows :

$$B = C\,K'\,[1 + \tfrac{1}{2}\,(R' - P) \sin 1'' \sin H]\,,\cdot$$

$$D = \frac{A}{K'}\,[1 - \tfrac{1}{2}\,(p - r')\,K' \sin 1'' \sin h]\,.$$

By means of Logs. A, B, C, and D, the first correction of the distance is found by the following formulas : —

if

$$d = \text{apparent distance of } \mathbb{C} \text{ from } \odot \text{ or } *,$$
$$\Delta_1\, d = \text{first correction,}$$
$$A' = \quad A\,(p - r')\sin h \cot d,$$
$$B' = -\,B\,(p - r')\sin H \operatorname{cosec} d,$$
$$C' = -\,C\,(R' - P)\sin H \cot d,$$
$$D' = \quad D\,(R' - P)\sin h \operatorname{cosec} d,$$

then

$$\Delta_1\, d = (A' + B') + (C' + D'),$$

and in the directions for using the tables, $A' + B'$ is called the "whole correction for $\mathbb{C}$," and $C' + D'$ the "whole correction for $\odot$ or $*$."

TABLE VI. — *Second Correction of the Lunar Distance*, computed by the formula

$$\Delta_2\, d = -\tfrac{1}{2}\,\Delta_1\, d^2 \sin 1'' \cot d.$$

Strictly, this formula should be

$$\Delta_2\, d = -\tfrac{1}{2}\,\Delta\, d^2 \sin 1'' \cot d,$$

where $\Delta\, d = $ whole correction of distance $= \Delta_1\, d + \Delta_2\, d$, so that by entering the table first with $\Delta_1\, d$ we find only an approximate value of $\Delta_2\, d$. But if with this approximate value we form the approximate whole correction $\Delta_1\, d + \Delta_2\, d$, and enter with this as the argument in the place of $\Delta_1\, d$, we have the true value of $\Delta_2\, d$. It is evident, however, from the table itself, that, in most practical cases, this degree of precision is unnecessary.

TABLE VII. — *For finding the Correction of the Lunar Distance for the Contraction of the Moon's Semidiameter* (by refraction), computed by the formula

$$\Delta\, s = \Delta\, s_0 \frac{(A' + B')^2}{(p - r')^2 \cos^2 h}$$

where the notation already employed is preserved, and

$\Delta\, s_0 = $ contraction of the moon's vertical semidiameter at the altitude h,

$\Delta\, s = $ contraction of the inclined semidiameter, or of that which lies in the direction of the lunar distance.

This table is subdivided into Tables VII. A. and VII. B. If we put

$$g = \frac{\Delta\, s_0}{(p - r')^2 \cos^2 h} \times f,$$

then

$$\Delta\, s = (A' + B')^2 \times \frac{g}{f},$$

where f is an arbitrary factor employed to give g convenient integral values, and in these tables we have taken $f = 18000000$; $\Delta\, s_0$, $\Delta\, s$, $p - r'$, and $A' + B'$ being all expressed in seconds. Table VII. A. gives the value of g with the arguments $p - r'$ and h; and Table VII. B. gives $\Delta\, s$ with the arguments $A' + B'$ and g.

To find $\Delta\, s_0$ from the arguments $p - r'$ and h, the *mean* value of r' for the altitude h was added to $p - r'$, so that p became known; whence also the semidiameter, —

and consequently, by .neans of the refraction tables, the contraction Δs_0. The value of Δs given by Table VII. B. is therefore that which corresponds to a mean state of the air, and in *extreme* cases may be in error 4″, but in no *probable* case will it be in error more than 1″. The true value of Δs might, however, always be found by correcting it, as refraction, by Tables IV. A. and IV. B.

TABLE VIII. — *For finding the Correction of the Lunar Distance for the Contraction of the Sun's Semidiameter* (by refraction), computed by the formula

$$\Delta S = \Delta S_0 \frac{(C' + D')^2}{(R' - P)^2 \cos^2 H},$$

where, in addition to the notation already employed, $\Delta S_0 =$ contraction of the sun's vertical semidiameter at the altitude H; $\Delta S =$ contraction of the inclined semidiameter. Table VIII. A., with the arguments $R' - P$ and H, gives a number

$$G = \frac{(R' - P)^2 \cos^2 H}{\Delta S_0} \times F,$$

in which ΔS_0 is taken for a mean value of the sun's semidiameter, P is assumed at 8″.5, and F is an arbitrary factor $= \frac{1}{200}$; $R' - P$, $C' + D'$, ΔS_0, and ΔS being all expressed in seconds. Table VIII. B., with the arguments $C' + D'$ and G, gives

$$\Delta S = (C' + D')^2 \times \frac{F}{G}.$$

This value of ΔS is that which belongs to the actual state of the air, but is for a mean value of the sun's semidiameter, the variations of which would not change the tabular value by more than 0″.5 in any case.

TABLE IX. — *Logarithms of Seconds.* This table contains the common logarithm answering to the arc (either space or time) in the argument. For the convenience of the navigator, this table is given with the argument, degrees, minutes, and seconds, or hours, minutes, and seconds, for every ten seconds, at the side, the unit figure of the seconds being found at the top.

The logarithm is given with the proper characteristic prefixed.

TABLE X. — This table contains the correction for second differences of the moon's motion.

TABLE XI. contains the values of log. N for the distances from the sun, computed by the following formula : —

$$N = N' \pi \left(\frac{\sin \Delta}{\sin d} - \frac{\sin \delta}{\tan d} \right);$$

where $\pi = $ ☾'s equatorial horizontal parallax,

$\Delta = $ sun's declination,

$\delta = $ ☾'s declination,

$d = $ angular distance of the moon and sun referred to the centre of the earth,

$N' = \dfrac{\epsilon^2}{\sqrt{(1 - \epsilon^2 \sin^2 \phi)}}$, for which the value of N' is taken corresponding to

$\phi = 45°$, or log. $N' = 7.8170$.

16

TABLE XII. — *For finding the Value of N.* By this table an approximate value of N may be readily found, with sufficient accuracy for ordinary purposes. When very great accuracy is desired, the formula given in the explanation of Table XI. should be used. In Table XII. the moon's horizontal parallax is assumed at its mean value $= 57' \ 30''$, and the two parts of N are separately tabulated by the formulas

$$a \ (= \text{first part of } N) \ = \ - N' \ \pi \sin \delta \cot an \ d,$$

$$b \ (= \text{second part of } N) = \ N' \ \pi \sin \Delta \csc \ d ;$$

from which N is found by the formula

$$N = a + b.$$

NOTE. — The demonstration of most of the formulas cited in the preceding explanation is given in the *Astronomical Journal* (Cambridge, Mass.), Vol. II., Nos. 3 and 4.

TABLES.

TABLE I.

Dip of the Sea Horizon.

Height of the Eye.	Dip of the Horizon.
Feet.	′ ″
0	0 00
1	0 59
2	1 23
3	1 42
4	1 58
5	2 11
6	2 24
7	2 36
8	2 46
9	2 56
10	3 06
11	3 15
12	3 24
13	3 32
14	3 40
15	3 48
16	3 55
17	4 02
18	4 09
19	4 16
20	4 23
21	4 29
22	4 36
23	4 42
24	4 48
25	4 54
26	5 00
27	5 06
28	5 11
29	5 17
30	5 22
35	5 48
40	6 12
45	6 34
50	6 56
55	7 16
60	7 35
65	7 54
70	8 12
75	8 29
80	8 46
85	9 02
90	9 18
95	9 33
100	9 48

TABLE II.

Augmentation of the Moon's Semidiameter.

Apparent Altitude of ☽.	☽'s Semidiameter.					
	14′	15′		16′		17′
°	30″	0″	30″	0″	30″	0″
0	0.1	0.1	0.1	0.1	0.2	0.2
2	0.6	0.6	0.7	0.7	0.8	0.8
4	1.0	1.1	1.2	1.3	1.4	1.5
6	1.5	1.6	1.7	1.9	2.0	2.1
8	2.0	2.1	2.3	2.4	2.6	2.7
10	2.4	2.6	2.8	3.0	3.2	3.4
12	2.9	3.1	3.3	3.6	3.8	4.0
14	3.4	3.6	3.9	4.1	4.4	4.7
16	3.8	4.1	4.4	4.7	5.0	5.3
18	4.3	4.6	4.9	5.2	5.6	5.9
20	4.7	5.1	5.4	5.8	6.1	6.5
22	5.2	5.5	5.9	6.3	6.7	7.1
24	5.6	6.0	6.4	6.8	7.3	7.7
26	6.0	6.5	6.9	7.4	7.8	8.3
28	6.5	6.9	7.4	7.9	8.4	8.9
30	6.9	7.3	7.9	8.4	8.9	9.5
32	7.3	7.8	8.3	8.9	9.4	10.0
34	7.7	8.2	8.8	9.4	10.0	10.6
36	8.1	8.6	9.2	9.8	10.5	11.1
38	8.4	9.0	9.7	10.3	10.9	11.6
40	8.8	9.4	10.1	10.7	11.4	12.1
42	9.2	9.8	10.5	11.2	11.9	12.6
44	9.5	10.2	10.9	11.6	12.3	13.1
46	9.8	10.5	11.3	12.0	12.8	13.6
48	10.2	10.9	11.6	12.4	13.2	14.0
50	10.5	11.2	12.0	12.8	13.6	14.4
52	10.8	11.5	12.3	13.1	14.0	14.9
54	11.1	11.8	12.7	13.5	14.4	15.3
56	11.3	12.1	13.0	13.8	14.7	15.6
58	11.6	12.4	13.3	14.1	15.1	16.0
60	11.8	12.7	13.5	14.4	15.4	16.3
62	12.1	12.9	13.8	14.7	15.7	16.6
64	12.3	13.2	14.1	15.0	16.0	16.9
66	12.5	13.4	14.3	15.2	16.2	17.2
68	12.7	13.6	14.5	15.5	16.5	17.5
70	12.9	13.8	14.7	15.7	16.7	17.7
72	13.0	13.9	14.9	15.9	16.9	17.9
74	13.1	14.1	15.0	16.0	17.1	18.1
76	13.3	14.2	15.2	16.2	17.2	18.3
78	13.4	14.3	15.3	16.3	17.4	18.4
80	13.5	14.4	15.4	16.4	17.5	18.6
82	13.5	14.5	15.5	16.5	17.6	18.7
84	13.6	14.6	15.6	16.6	17.6	18.7
86	13.6	14.6	15.6	16.6	17.7	18.8
88	13.7	14.6	15.6	16.7	17.7	18.8
90	13.7	14.6	15.6	16.7	17.7	18.8

TABLE III.

Augmentation of the Moon's Hor. Parallax.

Latitude of Observation.	☽'s Horizontal Parallax.		
	53′	57′	61′
°	″	′	″
0	0.0	0.0	0.0
2	0.0	0.0	0.0
4	0.1	0.1	0.1
6	0.1	0.1	0.1
8	0.2	0.2	0.2
10	0.3	0.3	0.4
12	0.5	0.5	0.5
14	0.6	0.7	0.7
16	0.8	0.9	0.9
18	1.0	1.1	1.1
20	1.2	1.3	1.4
22	1.5	1.6	1.7
24	1.7	1.9	2.0
26	2.0	2.2	2.3
28	2.3	2.5	2.6
30	2.6	2.8	3.0
32	2.9	3.1	3.4
34	3.3	3.5	3.8
36	3.6	3.9	4.1
38	4.0	4.3	4.6
40	4.3	4.6	5.0
42	4.7	5.0	5.4
44	5.0	5.4	5.8
46	5.4	5.8	6.2
48	5.8	6.2	6.6
50	6.1	6.6	7.1
52	6.5	7.0	7.5
54	6.8	7.4	7.9
56	7.2	7.7	8.3
58	7.5	8.1	8.6
60	7.8	8.4	9.0
62	8.1	8.8	9.4
64	8.4	9.1	9.7
66	8.7	9.4	10.0
68	9.0	9.7	10.3
70	9.2	9.9	10.6
72	9.5	10.2	10.9
74	9.7	10.4	11.1
76	9.8	10.6	11.3
78	10.0	10.8	11.5
80	10.1	10.9	11.7
82	10.3	11.0	11.8
84	10.3	11.1	11.9
86	10.4	11.2	12.0
88	10.4	11.2	12.0
90	10.5	11.3	12.0

TABLE IV.

Mean *Reduced Refraction* for Lunars. Barometer 30 inches. Fahrenheit's Thermometer 50°.

Apparent Altitude	Reduced Refraction	Diff. to 1'.	Apparent Altitude	Reduced Refraction	Apparent Altitude	Reduced Refraction	Apparent Altitude	Reduced Refraction
° '	' "	"	° '	' "	° '	' "	° '	' "
5 0	9 54.2	1.6	10 0	5 24.1	15 0	3 41.7	27 0	2 7.8
5	9 46.3	1.5	5	5 21.6	10	3 39.4	27 30	2 5.7
10	9 38.6	1.5	10	5 19.2	20	3 37.1	28 0	2 3.7
15	9 31.0	1.5	15	5 16.8	30	3 34.9	28 30	2 1.7
20	9 23.7	1.4	20	5 14.4	40	3 32.7	29 0	1 59.8
25	9 16.5	1.4	25	5 12.1	50	3 30.6	29 30	1 58.0
5 30	9 9.5	1.4	10 30	5 9.8	16 0	3 28.5	30 0	1 56.2
35	9 2.7	1.3	35	5 7.5	10	3 26.5	30 30	1 54.5
40	8 56.0	1.3	40	5 5.3	20	3 24.5	31 0	1 52.8
45	8 49.5	1.3	45	5 3.1	30	3 22.6	31 30	1 51.2
50	8 43.1	1.2	50	5 0.9	40	3 20.7	32 0	1 49.7
55	8 36.9	1.2	55	4 58.8	50	3 18.8	32 30	1 48.2
6 0	8 30.9	1.2	11 0	4 56.7	17 0	3 16.9	33 0	1 46.7
5	8 24.9	1.2	5	4 54.6	10	3 15.1	33 30	1 45.3
10	8 19.1	1.1	10	4 52.5	20	3 13.4	34 0	1 44.0
15	8 13.4	1.1	15	4 50.5	30	3 11.6	34 30	1 42.7
20	8 7.8	1.1	20	4 48.5	40	3 9.9	35 0	1 41.4
25	8 2.4	1.1	25	4 46.6	50	3 8.2	35 30	1 40.2
6 30	7 57.0	1.0	11 30	4 44.6	18 0	3 6.6	36 0	1 39.0
35	7 51.8	1.0	35	4 42.7	10	3 5.0	36 30	1 37.8
40	7 46.7	1.0	40	4 40.8	20	3 3.4	37 0	1 36.7
45	7 41.7	1.0	45	4 38.9	30	3 1.8	37 30	1 35.6
50	7 36.7	1.0	50	4 37.1	40	3 0.3	38 0	1 34.5
55	7 31.9	0.9	55	4 35.3	50	2 58.8	38 30	1 33.5
7 0	7 27.2	0.9	12 0	4 33.5	19 0	2 57.3	39 0	1 32.5
5	7 22.6	0.9	5	4 31.7	10	2 55.9	39 30	1 31.5
10	7 18.0	0.9	10	4 30.0	20	2 54.4	40 0	1 30.6
15	7 13.6	0.9	15	4 28.3	30	2 53.0	40 30	1 29.6
20	7 9.2	0.9	20	4 26.6	40	2 51.6	41 0	1 28.7
25	7 4.9	0.8	25	4 24.9	50	2 50.3	41 30	1 27.8
7 30	7 0.8	0.8	12 30	4 23.2	20 0	2 49.0	42 0	1 27.0
35	6 56.6	0.8	35	4 21.6	10	2 47.6	42 30	1 26.2
40	6 52.6	0.8	40	4 20.0	20	2 46.4	43 0	1 25.4
45	6 48.6	0.8	45	4 18.4	30	2 45.1	43 30	1 24.6
50	6 44.8	0.8	50	4 16.8	40	2 43.8	44 0	1 23.8
55	6 40.9	0.7	55	4 15.2	50	2 42.6	44 30	1 23.1
8 0	6 37.2	0.7	13 0	4 13.7	21 0	2 41.4	45 0	1 22.4
5	6 33.5	0.7	5	4 12.2	10	2 40.2	46 0	1 21.0
10	6 29.9	0.7	10	4 10.7	20	2 39.0	47 0	1 19.6
15	6 26.3	0.7	15	4 9.2	30	2 37.9	48 0	1 18.4
20	6 22.8	0.7	20	4 7.7	40	2 36.7	49 0	1 17.2
25	6 19.4	0.7	25	4 6.3	50	2 35.6	50 0	1 16.0
8 30	6 16.0	0.7	13 30	4 4.8	22 0	2 34.5	51 0	1 15.0
35	6 12.7	0.6	35	4 3.4	10	2 33.4	52 0	1 13.9
40	6 9.5	0.6	40	4 2.0	20	2 32.4	53 0	1 13.0
45	6 6.3	0.6	45	4 0.6	30	2 31.3	54 0	1 12.0
50	6 3.1	0.6	50	3 59.3	40	2 30.3	55 0	1 11.1
55	6 0.0	0.6	55	3 57.9	50	2 29.2	56 0	1 10.3
9 0	5 57.0	0.6	14 0	3 56.6	23 0	2 28.2	57 0	1 9.5
5	5 54.0	0.6	5	3 55.3	20	2 26.3	58 0	1 8.7
10	5 51.1	0.6	10	3 54.0	40	2 24.4	59 0	1 8.0
15	5 48.2	0.6	15	3 52.7	24 0	2 22.5	60 0	1 7.3
20	5 45.3	0.6	20	3 51.4	20	2 20.7	62 0	1 6.0
25	5 42.5	0.5	25	3 50.1	40	2 18.9	64 0	1 4.9
9 30	5 39.8	0.5	14 30	3 48.9	25 0	2 17.2	66 0	1 3.8
35	5 37.0	0.5	35	3 47.6	20	2 15.5	68 0	1 2.9
40	5 34.4	0.5	40	3 46.4	40	2 13.9	70 0	1 2.0
45	5 31.7	0.5	45	3 45.2	26 0	2 12.3	73 0	1 1.0
50	5 29.2	0.5	50	3 44.0	20	2 10.8	76 0	1 0.1
55	5 26.6	0.5	55	3 42.8	40	2 9.3	80 0	0 59.2
10 0	5 24.1		15 0	3 41.7	27 0	2 7.8	90 0	0 58.3

TABLE IV. A.

Correction of the Mean Refraction for the Height of the Barometer.

Barometer. Subtract.	0'	1'	2'	3'	4'	5'	6'	7'	8'	9'	10'	Barometer. Add.
	0″ 30″	0″ 30″	0″ 30″	0″ 30″	0″ 30″	0″ 30″	0″ 30″	0″ 30″	0″ 30″	0″ 30″	0″	
27.50	0 2	5 7	10 12	15 17	20 23	25 28	30 33	35 38	40 43	45 48	51	
27.55	0 2	5 7	10 12	15 17	20 22	25 27	30 32	35 37	40 42	45 47	50	
27.60	0 2	5 7	10 12	14 17	19 22	24 27	29 31	34 36	39 41	44 46	49	
27.65	0 2	5 7	9 12	14 16	19 21	24 26	28 31	33 36	38 40	43 45	48	
27.70	0 2	5 7	9 11	14 16	18 21	23 25	28 30	32 35	37 39	42 44	47	
27.75	0 2	4 7	9 11	13 16	18 20	23 25	27 29	32 34	36 39	41 43	46	
27.80	0 2	4 7	9 11	13 15	18 20	22 24	27 29	31 33	35 38	40 42	45	
27.85	0 2	4 6	9 11	13 15	17 19	22 24	26 28	30 32	35 37	39 41	44	
27.90	0 2	4 6	8 10	13 15	17 19	21 23	25 27	30 32	34 36	38 40	43	
27.95	0 2	4 6	8 10	12 14	16 18	21 23	25 27	29 31	33 35	37 39	42	
28.00	0 2	4 6	8 10	12 14	16 18	20 22	24 26	28 30	32 34	36 38	41	
28.05	0 2	4 6	8 10	12 14	16 18	20 22	24 25	27 29	31 33	35 37	39	
28.10	0 2	4 6	8 9	11 13	15 17	19 21	23 25	27 29	31 33	34 36	38	
28.15	0 2	4 6	7 9	11 13	15 17	19 20	22 24	26 28	30 32	34 36	37	
28.20	0 2	4 5	7 9	11 13	14 16	18 20	22 24	25 27	29 31	33 35	36	
28.25	0 2	3 5	7 9	10 12	14 16	18 19	21 23	25 26	28 30	32 34	35	
28.30	0 2	3 5	7 8	10 12	14 15	17 19	21 22	24 26	27 29	31 33	34	
28.35	0 2	3 5	7 8	10 12	13 15	17 18	20 22	23 25	27 28	30 32	33	
28.40	0 2	3 5	6 8	10 11	13 14	16 18	19 21	23 24	26 27	29 31	32	
28.45	0 2	3 5	6 8	9 11	12 14	16 17	19 20	22 23	25 27	28 30	31	
28.50	0 1	3 4	6 7	9 10	12 14	15 17	18 20	21 23	24 26	27 29	30	31.50
28.55	0 1	3 4	6 7	9 10	12 13	15 16	17 19	20 22	23 25	26 28	29	31.45
28.60	0 1	3 4	6 7	8 10	11 13	14 15	17 18	20 21	23 24	25 27	28	31.40
28.65	0 1	3 4	5 7	8 9	11 12	14 15	16 18	19 20	22 23	25 26	27	31.35
28.70	0 1	3 4	5 6	8 9	10 12	13 14	16 17	18 20	21 22	24 25	26	31.30
28.75	0 1	2 4	5 6	7 9	10 11	13 14	15 16	18 19	20 21	23 24	25	31.25
28.80	0 1	2 4	5 6	7 8	10 11	12 13	14 16	17 18	19 21	22 23	24	31.20
28.85	0 1	2 3	5 6	7 8	9 10	12 13	14 15	16 17	19 20	21 22	23	31.15
28.90	0 1	2 3	4 5	7 8	9 10	11 12	13 14	16 17	18 19	20 21	22	31.10
28.95	0 1	2 3	4 5	6 7	8 9	11 12	13 14	15 16	17 18	19 20	21	31.05
29.00	0 1	2 3	4 5	6 7	8 9	10 11	12 13	14 15	16 17	18 19	20	31.00
29.05	0 1	2 3	4 5	6 7	8 9	10 11	11 12	13 14	15 16	16 17	18	30.95
29.10	0 1	2 3	4 4	5 6	7 8	9 10	11 12	13 14	15 15	16 17	18	30.90
29.15	0 1	2 3	3 4	5 6	7 8	9 9	10 11	12 13	14 15	15 16	17	30.85
29.20	0 1	2 2	3 4	5 6	6 7	8 9	10 10	11 12	13 14	15 15	16	30.80
29.25	0 1	1 2	3 4	4 5	6 7	8 8	9 10	11 11	12 13	14 14	15	30.75
29.30	0 1	1 2	3 3	4 5	6 6	7 8	8 9	10 11	11 12	13 13	14	30.70
29.35	0 1	1 2	3 3	4 5	5 6	7 7	8 9	9 10	10 11	12 13	13	30.65
29.40	0 1	1 2	2 3	4 4	5 5	6 7	7 8	8 9	10 10	11 12	12	30.60
29.45	0 1	1 2	2 3	3 4	4 5	6 6	7 7	8 8	9 9	10 11	11	30.55
29.50	0 0	1 1	2 2	3 3	4 5	5 6	6 7	7 8	8 9	9 10	10	30.50
29.55	0 0	1 1	2 2	3 3	4 4	5 5	5 6	6 7	7 8	8 9	9	30.45
29.60	0 0	1 1	2 2	2 3	3 4	4 4	5 5	6 6	6 7	7 8	8	30.40
29.65	0 0	1 1	1 2	2 2	3 3	4 4	4 5	5 5	6 6	6 7	7	30.35
29.70	0 0	1 1	1 1	2 2	2 3	3 3	4 4	4 5	5 5	6 6	6	30.30
29.75	0 0	0 1	1 1	1 2	2 2	3 3	3 3	4 4	4 4	5 5	5	30.25
29.80	0 0	0 1	1 1	1 1	2 2	2 2	2 3	3 3	3 3	4 4	4	30.20
29.85	0 0	0 0	1 1	1 1	1 1	2 2	2 2	2 2	2 2	3 3	3	30.15
29.90	0 0	0 0	0 0	1 1	1 1	1 1	1 1	1 2	2 2	2 2	2	30.10
29.95	0 0	0 0	0 0	0 0	0 0	1 1	1 1	1 1	1 1	1 1	1	30.05
30.00	0 0	0 0	0 0	0 0	00 0	0 0	0 0	0 0	0 0	0 0	0	30.00
Subtract.	0″ 30″	0″ 30″	0″ 30″	0″ 30″	0″ 30″	0″ 30″	0″ 30″	0″ 30″	0″ 30″	0″ 30″	0″	Add.
Barometer.	0'	1'	2'	3'	4'	5'	6'	7'	8'	9'	10'	Barometer.

MEAN REFRACTION.

TABLE IV. B.

Correction of the Mean Refraction for the Height of the Thermometer.

MEAN REFRACTION.

Thermom. Add.	0′ 0″	0′ 30″	1′ 0″	1′ 30″	2′ 0″	2′ 30″	3′ 0″	3′ 30″	4′ 0″	4′ 30″	5′ 0″	5′ 30″	6′ 0″	6′ 30″	7′ 0″	7′ 30″	8′ 0″	8′ 30″	9′ 0″	9′ 30″	10′ 0″	Thermom. Add.
−10	0	4	8	12	16	20	24	28	33	37	41	46	50	55	60	65	70	75	80	85	90	−10
− 8	0	4	8	12	15	19	23	27	31	36	40	44	48	53	58	62	67	72	77	82	87	− 8
− 6	0	4	7	11	15	19	22	26	30	34	38	42	47	51	55	60	64	69	74	79	84	− 6
− 4	0	4	7	11	14	18	22	25	29	33	37	41	45	49	53	57	62	66	71	76	80	− 4
− 2	0	3	7	10	14	17	21	24	28	31	35	39	43	47	51	55	59	64	68	72	77	− 2
0	0	3	7	10	13	16	20	23	27	30	34	37	41	45	49	53	57	61	65	69	74	0
2	0	3	6	9	12	16	19	22	25	29	32	36	39	43	47	50	54	58	62	66	70	2
4	0	3	6	9	12	15	18	21	24	28	31	34	37	41	44	48	52	55	59	63	67	4
6	0	3	6	8	11	14	17	20	23	26	29	32	36	39	42	46	49	53	56	60	64	6
8	0	3	5	8	11	14	16	19	22	25	28	31	34	37	40	43	47	50	54	57	61	8
10	0	3	5	8	10	13	15	18	21	24	26	29	32	35	38	41	44	48	51	54	58	10
11	0	2	5	7	10	13	15	18	20	23	26	28	31	34	37	40	43	46	49	53	56	11
12	0	2	5	7	10	12	15	17	20	22	25	28	30	33	36	39	42	45	48	51	54	12
13	0	2	5	7	9	12	14	17	19	22	24	27	30	32	35	38	41	44	47	50	53	13
14	0	2	5	7	9	11	14	16	19	21	24	26	29	31	34	37	40	42	45	48	51	14
15	0	2	4	7	9	11	13	16	18	20	23	25	28	30	33	36	38	41	44	47	50	15
16	0	2	4	6	9	11	13	15	18	20	22	25	27	29	32	35	37	40	43	45	48	16
17	0	2	4	6	8	10	13	15	17	19	21	24	26	29	31	33	36	39	41	44	47	17
18	0	2	4	6	8	10	12	14	16	19	21	23	25	28	30	32	35	37	40	43	45	18
19	0	2	4	6	8	10	12	14	16	18	20	22	24	27	29	31	34	36	39	41	44	19
20	0	2	4	6	8	9	11	13	15	17	19	22	24	26	28	30	33	35	37	40	42	20
21	0	2	4	5	7	9	11	13	15	17	19	21	23	25	27	29	31	34	36	38	41	21
22	0	2	3	5	7	9	11	12	14	16	18	20	22	24	26	28	30	32	35	37	39	22
23	0	2	3	5	7	8	10	12	14	15	17	19	21	23	25	27	29	31	33	36	38	23
24	0	2	3	5	6	8	10	11	13	15	17	18	20	22	24	26	28	30	32	34	36	24
25	0	2	3	5	6	8	9	11	13	14	16	18	19	21	23	25	27	29	31	33	35	25
26	0	1	3	4	6	7	9	11	12	14	15	17	19	20	22	24	26	28	29	31	33	26
27	0	1	3	4	6	7	9	10	12	13	15	16	18	19	21	23	25	26	28	30	32	27
28	0	1	3	4	5	7	8	10	11	12	14	15	17	19	20	22	23	25	27	29	30	28
29	0	1	3	4	5	6	8	9	11	12	13	15	16	18	19	21	22	24	26	27	29	29
30	0	1	2	4	5	6	7	9	10	11	13	14	15	17	18	20	21	23	24	26	28	30
31	0	1	2	3	5	6	7	8	9	11	12	13	15	16	17	19	20	22	23	25	26	31
32	0	1	2	3	4	6	7	8	9	10	11	13	14	15	16	18	19	20	22	23	25	32
33	0	1	2	3	4	5	6	7	8	10	11	12	13	14	15	17	18	19	21	22	23	33
34	0	1	2	3	4	5	6	7	8	9	10	11	12	13	14	16	17	18	19	21	22	34
35	0	1	2	3	4	5	6	6	7	8	9	10	11	13	14	15	16	17	18	19	20	35
36	0	1	2	3	3	4	5	6	7	8	9	10	11	12	13	14	15	16	17	18	19	36
37	0	1	2	2	3	4	4	5	6	7	8	9	10	11	12	13	14	15	16	17	18	37
38	0	1	1	2	3	4	4	5	6	7	7	8	9	10	11	12	13	13	14	15	16	38
39	0	1	1	2	3	3	4	5	5	6	7	8	8	9	10	11	11	12	13	14	15	39
40	0	1	1	2	2	3	4	4	5	6	6	7	8	8	9	10	10	11	12	13	13	40
41	0	1	1	2	2	3	3	4	4	5	6	6	7	7	8	9	9	10	11	11	12	41
42	0	0	1	1	2	2	3	3	4	4	5	5	6	7	7	8	8	9	9	10	11	42
43	0	0	1	1	2	2	3	3	3	4	4	5	6	6	6	7	7	8	8	9	9	43
44	0	0	1	1	1	2	2	3	3	3	4	4	4	5	5	6	6	7	7	8	8	44
45	0	0	1	1	1	1	2	2	2	3	3	3	4	4	4	5	5	6	6	6	7	45
46	0	0	0	1	1	1	1	2	2	2	3	3	3	4	4	4	4	5	5	5	5	46
47	0	0	0	1	1	1	1	1	1	2	2	2	2	3	3	3	3	4	4	4	4	47
48	0	0	0	0	0	1	1	1	1	1	1	1	1	2	2	2	2	2	2	2	3	48
49	0	0	0	0	0	0	0	0	0	1	1	1	1	1	1	1	1	1	1	1	1	49
50	0	0	0	0	0	0	0	0	0	0	0	0	0	0	0	0	0	0	0	0	0	50

Add.	0″	30″	0″	30″	0″	30″	0″	30″	0″	30″	0″	30″	0″	30″	0″	30″	0″	30″	0″	30″	0″	Add.
Thermom.	0′		1′		2′		3′		4′		5′		6′		7′		8′		9′		10′	Thermom.

MEAN REFRACTION.

TABLE IV. B.

Correction of the Mean Refraction for the Height of the Thermometer.

MEAN REFRACTION.

Thermom. Subtract	0'		1'		2'		3'		4'		5'		6'		7'		8'		9'		10'	Thermom. Subtract
	´	30	´	30	´	30	´	30	´	30	´	30	´	30	´	30	´	30	´	30	´	
50°	0	0	0	0	0	0	0	0	0	0	0	0	0	0	0	0	0	0	0	0	0	50°
51	0	0	0	0	0	0	0	0	0	1	1	1	1	1	1	1	1	1	1	1	1	51
52	0	0	0	0	0	1	1	1	1	1	1	1	1	2	2	2	2	2	2	2	3	52
53	0	0	0	1	1	1	1	1	1	2	2	2	2	2	2	3	3	3	3	4	4	53
54	0	0	0	1	1	1	1	2	2	2	2	2	3	3	3	4	4	4	5	5	5	54
55	0	0	1	1	1	1	2	2	2	3	3	3	4	4	4	5	5	5	6	6	6	55
56	0	0	1	1	1	2	2	2	3	3	4	4	4	5	5	6	6	6	7	7	8	56
57	0	0	1	1	2	2	2	3	3	4	4	5	5	6	6	6	7	8	8	8	9	57
58	0	0	1	1	2	2	3	3	4	4	5	5	6	6	7	7	8	9	9	10	10	58
59	0	1	1	2	2	3	3	4	4	5	5	6	6	7	8	8	9	10	10	11	12	59
60	0	1	1	2	2	3	3	4	5	5	6	7	7	8	9	9	10	11	11	12	13	60
61	0	1	1	2	3	3	4	4	5	6	7	7	8	9	9	10	11	12	12	13	14	61
62	0	1	1	2	3	3	4	5	6	6	7	8	9	9	10	11	12	13	14	15	15	62
63	0	1	1	2	3	4	5	5	6	7	8	8	9	10	11	12	13	14	15	16	17	63
64	0	1	2	2	3	4	5	6	7	7	8	9	10	11	12	13	14	15	16	17	18	64
65	0	1	2	3	3	4	5	6	7	8	9	10	11	12	13	14	15	16	17	18	19	65
66	0	1	2	3	4	5	6	6	7	8	9	10	11	12	14	15	16	17	18	19	20	66
67	0	1	2	3	4	5	6	7	8	9	10	11	12	13	14	16	17	18	19	20	22	67
68	0	1	2	3	4	5	6	7	8	9	11	11	13	14	15	16	18	19	20	22	23	68
69	0	1	2	3	4	5	7	8	9	10	11	12	13	15	16	17	19	20	21	23	24	69
70	0	1	2	3	5	6	7	8	9	10	12	12	14	16	17	18	20	21	22	24	25	70
71	0	1	2	4	5	6	7	8	10	11	12	13	15	16	18	19	20	22	23	25	27	71
72	0	1	2	4	5	6	8	9	10	11	13	14	16	17	18	20	21	23	25	26	28	72
73	0	1	3	4	5	7	8	9	11	12	13	14	16	18	19	21	22	24	26	27	29	73
74	0	1	3	4	5	7	8	10	11	12	14	15	17	18	20	22	23	25	27	28	30	74
75	0	1	3	4	6	7	8	10	11	13	14	16	18	19	21	22	24	26	28	29	31	75
76	0	1	3	4	6	7	9	10	12	13	15	16	18	20	22	23	25	27	29	31	32	76
77	0	1	3	5	6	8	9	11	12	14	16	17	19	21	22	24	26	28	30	32	34	77
78	0	2	3	5	6	8	9	11	13	14	16	18	20	21	23	25	27	29	31	33	35	78
79	0	2	3	5	6	8	10	11	13	15	17	19	20	22	24	26	28	30	32	34	36	79
80	0	2	3	5	7	8	10	12	14	15	17	19	21	23	25	27	29	31	33	35	37	80
81	0	2	3	5	7	9	10	12	14	16	18	20	21	24	26	28	30	32	34	36	38	81
82	0	2	4	5	7	9	11	13	14	16	18	20	22	24	26	28	31	33	35	37	40	82
83	0	2	4	5	7	9	11	13	15	17	19	21	23	25	27	29	31	34	36	38	41	83
84	0	2	4	6	8	9	11	13	15	17	19	21	23	26	28	30	32	35	37	39	42	84
85	0	2	4	6	8	10	12	14	16	18	20	22	24	26	29	31	33	36	38	40	43	85
86	0	2	4	6	8	10	12	14	16	18	20	23	25	27	29	32	34	37	39	42	44	86
87	0	2	4	6	8	10	12	14	17	19	21	23	25	28	30	32	35	38	40	43	45	87
88	0	2	4	6	8	10	13	15	17	19	21	24	26	28	31	33	36	38	41	44	46	88
89	0	2	4	6	9	11	13	15	17	20	22	24	27	29	32	34	37	39	42	45	48	89
90	0	2	4	7	9	11	13	16	18	20	23	25	27	30	32	35	38	40	43	46	49	90
91	0	2	4	7	9	11	14	16	18	21	23	25	28	31	33	36	39	41	44	47	50	91
92	0	2	5	7	9	11	14	16	19	21	24	26	29	31	34	37	39	42	45	48	51	92
93	0	2	5	7	9	12	14	17	19	22	24	27	29	32	35	37	40	43	46	49	52	93
94	0	2	5	7	10	12	14	17	19	22	25	27	30	33	35	38	41	44	47	50	53	94
95	0	2	5	7	10	12	15	17	20	22	25	28	30	33	36	39	42	45	48	51	54	95
96	0	2	5	7	10	12	15	18	20	23	26	28	31	34	37	40	43	46	49	52	55	96
97	0	3	5	8	10	13	15	18	21	23	26	29	32	35	38	41	44	47	50	53	56	97
98	0	3	5	8	10	13	16	18	21	24	27	29	32	35	38	41	44	48	51	54	58	98
99	0	3	5	8	11	13	16	19	21	24	27	30	33	36	39	42	45	49	52	55	59	99
100	0	3	5	8	11	13	16	19	22	25	28	31	34	37	40	43	46	50	53	56	60	100

| Subtract | ´ | 30 | ´ | 30 | ´ | 30 | ´ | 30 | ´ | 30 | ´ | 30 | ´ | 30 | ´ | 30 | ´ | 30 | ´ | 30 | ´ | Subtract |
| Thermom. | 0' | | 1' | | 2' | | 3' | | 4' | | 5' | | 6' | | 7' | | 8' | | 9' | | 10' | Thermom. |

MEAN REFRACTION.

TABLE V. LOG. A.

Logs A, B, C, and D, for Computing the *First Correction* of the Lunar Distance.

App. Alt. of D	41'	42'	43'	44'	45'	46'	47'	48'	49'	50'	51	52'	53'	54'	55
5° 0'	.0288	0295	0301	0308	0315	0321	0328	0335	0341	0348	0355	0361	0366		
2	.0286	0293	0299	0306	0313	0319	0326	0333	0339	0346	0352	0359	0366		
4	.0284	0291	0297	0304	0311	0317	0324	0330	0337	0344	0350	0357	0363		
6	.0282	0289	0296	0302	0309	0315	0322	0328	0335	0341	0348	0354	0361		
8	.0281	0287	0294	0300	0307	0313	0320	0326	0333	0339	0346	0352	0359		
5 10	.0279	0285	0292	0298	0305	0311	0318	0324	0331	0337	0344	0350	0356		
12	.0277	0284	0290	0296	0303	0309	0316	0322	0329	0335	0341	0348	0354		
14	.0275	0282	0288	0295	0301	0307	0314	0320	0327	0333	0339	0346	0352		
16	.0274	0280	0286	0293	0299	0306	0312	0318	0325	0331	0337	0344	0350		
18	.0272	0278	0285	0291	0297	0304	0310	0316	0323	0329	0335	0341	0348		
5 20	.0270	0277	0283	0289	0296	0302	0308	0314	0321	0327	0333	0339	0346		
22	.0269	0275	0281	0288	0294	0300	0306	0313	0319	0325	0331	0337	0344		
24	.0267	0273	0280	0286	0292	0298	0304	0311	0317	0323	0329	0335	0341		
26	.0265	0272	0278	0284	0290	0296	0303	0309	0315	0321	0327	0333	0339	0346	
28	.0264	0270	0276	0282	0289	0295	0301	0307	0313	0319	0325	0331	0337	0344	
5 30	.0262	0268	0275	0281	0287	0293	0299	0305	0311	0317	0323	0329	0335	0342	
32	.0261	0267	0273	0279	0285	0291	0297	0303	0309	0315	0321	0327	0334	0340	
34	.0259	0265	0271	0277	0283	0290	0296	0302	0308	0314	0320	0326	0332	0338	
36	.0258	0264	0270	0276	0282	0288	0294	0300	0306	0312	0318	0324	0330	0336	
38		0262	0268	0274	0280	0286	0292	0298	0304	0310	0316	0322	0328	0334	
5 40		0261	0267	0273	0279	0285	0290	0296	0302	0308	0314	0320	0326	0332	
42		0259	0265	0271	0277	0283	0289	0295	0301	0306	0312	0318	0324	0330	
44		0258	0264	0270	0275	0281	0287	0293	0299	0305	0311	0316	0322	0328	
46		0256	0262	0268	0274	0280	0286	0291	0297	0303	0309	0315	0320	0326	
48		0255	0261	0267	0272	0278	0284	0290	0296	0301	0307	0313	0319	0324	
5 50		0253	0259	0265	0271	0277	0282	0288	0294	0300	0305	0311	0317	0323	
52		0252	0258	0264	0269	0275	0281	0287	0292	0298	0304	0309	0315	0321	
54		0251	0256	0262	0268	0274	0279	0285	0291	0296	0302	0308	0313	0319	
56		0249	0255	0261	0266	0272	0278	0283	0289	0295	0300	0306	0312	0317	
58		0248	0254	0259	0265	0271	0276	0282	0287	0293	0299	0304	0310	0316	
6 0		0247	0252	0258	0263	0269	0275	0280	0286	0291	0297	0303	0308	0314	
2		0245	0251	0256	0262	0268	0273	0279	0284	0290	0295	0301	0307	0312	
4		0244	0249	0255	0261	0266	0272	0277	0283	0288	0294	0299	0305	0310	
6		0243	0248	0254	0259	0265	0270	0276	0281	0287	0292	0298	0303	0309	
8		0241	0247	0252	0258	0263	0269	0274	0280	0285	0291	0296	0302	0307	
6 10		0240	0246	0251	0256	0262	0267	0273	0278	0284	0289	0295	0300	0306	
12		0239	0244	0250	0255	0261	0266	0271	0277	0282	0288	0293	0299	0304	
14		0237	0243	0248	0254	0259	0265	0270	0275	0281	0286	0292	0297	0302	
16		0236	0242	0247	0252	0258	0263	0269	0274	0279	0285	0290	0295	0301	
18		0235	0240	0246	0251	0257	0262	0267	0273	0278	0283	0289	0294	0299	
6 20		0234	0239	0245	0250	0255	0261	0266	0271	0276	0282	0287	0292	0298	
22		0233	0238	0243	0249	0254	0259	0264	0270	0275	0280	0286	0291	0296	
24		0231	0237	0242	0247	0253	0258	0263	0268	0274	0279	0284	0289	0295	
26			0236	0241	0246	0251	0257	0262	0267	0272	0277	0283	0288	0293	
28			0234	0240	0245	0250	0255	0260	0266	0271	0276	0281	0286	0292	0297
6 30			0233	0238	0244	0249	0254	0259	0264	0270	0275	0280	0285	0290	0295
32			0232	0237	0242	0248	0253	0258	0263	0268	0273	0278	0284	0289	0294
34			0231	0236	0241	0246	0251	0257	0262	0267	0272	0277	0282	0287	0292
36			0230	0235	0240	0245	0250	0255	0260	0266	0271	0276	0281	0286	0291
38			0229	0234	0239	0244	0249	0254	0259	0264	0269	0274	0279	0284	0290
6 40			0227	0232	0238	0243	0248	0253	0258	0263	0268	0273	0278	0283	0288
42			0226	0231	0236	0241	0246	0252	0257	0262	0267	0272	0277	0282	0287
44			0225	0230	0235	0240	0245	0250	0255	0260	0265	0270	0275	0280	0285
46			0224	0229	0234	0239	0244	0249	0254	0259	0264	0269	0274	0279	0284
48			0223	0228	0233	0238	0243	0248	0253	0258	0263	0268	0273	0278	0283
6 50			0222	0227	0232	0237	0242	0247	0252	0257	0262	0266	0271	0276	0281
52			0221	0226	0231	0236	0241	0246	0250	0255	0260	0265	0270	0275	0280
54			0220	0225	0230	0235	0239	0244	0249	0254	0259	0263	0268	0274	0279
56			0219	0224	0229	0233	0238	0243	0248	0253	0258	0263	0267	0272	0277
58			0218	0223	0227	0232	0237	0242	0247	0252	0257	0261	0266	0271	0276
7 0			0217	0222	0226	0231	0236	0241	0246	0251	0255	0260	0265	0270	0275

TABLE V. LOG. A.

Logs. A, B, C, and D, for Lunars.

App. Alt. of ☾	REDUCED PARALLAX AND REFRACTION OF ☾													
	44'	45'	46'	47'	48'	49'	50'	51'	52'	53'	54'	55'	56'	57'
7° 0'	.0222	0226	0231	0236	0241	0246	0251	0255	0260	0265	0270	0275		
3	.0220	0225	0230	0234	0239	0244	0249	0254	0258	0263	0268	0273		
6	.0218	0223	0228	0233	0238	0242	0247	0252	0257	0261	0266	0271		
9	.0217	0222	0226	0231	0236	0241	0245	0250	0255	0260	0264	0269		
12	.0215	0220	0225	0230	0234	0239	0244	0248	0253	0258	0262	0267		
7 15	.0214	0219	0223	0228	0233	0237	0242	0247	0251	0256	0261	0265		
18	.0213	0217	0222	0226	0231	0236	0240	0245	0250	0254	0259	0263		
21	.0211	0216	0220	0225	0230	0234	0239	0243	0248	0253	0257	0262		
24	.0210	0214	0219	0223	0228	0233	0237	0242	0246	0251	0255	0260		
27	.0208	0213	0217	0222	0227	0231	0236	0240	0245	0249	0254	0258		
7 30	.0207	0211	0216	0220	0225	0230	0234	0239	0243	0248	0252	0257		
33	.0206	0210	0215	0219	0224	0228	0232	0237	0241	0246	0250	0255		
36	.0204	0209	0213	0218	0222	0227	0231	0235	0240	0244	0249	0253		
39	.0203	0207	0212	0216	0221	0225	0229	0234	0238	0243	0247	0252		
42	.0202	0206	0210	0215	0219	0224	0228	0232	0237	0241	0246	0250		
7 45	.0200	0205	0209	0213	0218	0222	0227	0231	0235	0240	0244	0248		
48	.0199	0203	0208	0212	0216	0221	0225	0229	0234	0238	0242	0247		
51	.0198	0202	0206	0211	0215	0219	0224	0228	0232	0237	0241	0245	0249	
54	.0196	0201	0205	0209	0214	0218	0222	0227	0231	0235	0239	0244	0248	
57	.0195	0200	0204	0208	0212	0217	0221	0225	0229	0234	0238	0242	0246	
8 0	.0194	0198	0203	0207	0211	0215	0219	0224	0228	0232	0236	0241	0245	
3	.0193	0197	0201	0206	0210	0214	0218	0222	0227	0231	0235	0239	0243	
6	.0192	0196	0200	0204	0208	0213	0217	0221	0225	0229	0233	0238	0242	
9		0195	0199	0203	0207	0211	0215	0220	0224	0228	0232	0236	0240	
12		0193	0198	0202	0206	0210	0214	0218	0222	0227	0231	0235	0239	
8 15		0192	0196	0201	0205	0209	0213	0217	0221	0225	0229	0233	0237	
18		0191	0195	0199	0203	0207	0212	0217	0220	0224	0228	0232	0236	
21		0190	0194	0198	0202	0206	0210	0214	0218	0222	0226	0231	0235	
24		0189	0193	0197	0201	0205	0209	0213	0217	0221	0225	0229	0233	
27		0188	0192	0196	0200	0204	0208	0212	0216	0220	0224	0228	0232	
8 30		0187	0191	0195	0199	0203	0207	0211	0215	0219	0223	0226	0230	
33		0186	0190	0193	0197	0201	0205	0209	0213	0217	0221	0225	0229	
36		0184	0188	0192	0196	0200	0204	0208	0212	0216	0220	0224	0228	
39		0183	0187	0191	0195	0199	0203	0207	0211	0215	0219	0223	0226	
42		0182	0186	0190	0194	0198	0202	0206	0210	0214	0217	0221	0225	
8 45		0181	0185	0189	0193	0197	0201	0205	0208	0212	0216	0220	0224	
48		0180	0184	0188	0192	0196	0200	0203	0207	0211	0215	0219	0223	
51		0179	0183	0187	0191	0195	0198	0202	0206	0210	0214	0218	0221	
54		0178	0182	0186	0190	0193	0197	0201	0205	0209	0212	0216	0220	
57		0177	0181	0185	0189	0192	0196	0200	0204	0208	0211	0215	0219	
9 0		0176	0180	0184	0188	0191	0195	0199	0203	0206	0210	0214	0218	
3		0175	0179	0183	0186	0190	0194	0198	0201	0205	0209	0213	0216	
6		0174	0178	0182	0185	0189	0193	0197	0200	0204	0208	0211	0215	
9		0173	0177	0181	0184	0188	0192	0196	0199	0203	0207	0210	0214	
12		0172	0176	0180	0183	0187	0191	0194	0198	0202	0206	0209	0213	
9 15		0171	0175	0179	0182	0186	0190	0193	0197	0201	0204	0208	0212	
18		0170	0174	0178	0181	0185	0189	0192	0196	0200	0203	0207	0211	
21		0170	0173	0177	0180	0184	0188	0191	0195	0199	0202	0206	0209	
24			0172	0176	0179	0183	0187	0190	0194	0198	0201	0205	0208	
27			0171	0175	0179	0182	0186	0189	0193	0196	0200	0204	0207	
9 30			0170	0174	0178	0181	0185	0188	0192	0195	0199	0203	0206	
33			0170	0173	0177	0180	0184	0187	0191	0194	0198	0201	0205	
36			0169	0172	0176	0179	0183	0186	0190	0193	0197	0200	0204	
39			0168	0171	0175	0178	0182	0185	0189	0192	0196	0199	0203	
42			0167	0170	0174	0177	0181	0184	0188	0191	0195	0198	0202	
9 45			0166	0169	0173	0176	0180	0183	0187	0190	0194	0197	0201	
48			0165	0169	0172	0176	0179	0182	0186	0189	0193	0196	0200	0203
51			0164	0168	0171	0175	0178	0182	0185	0188	0192	0195	0199	0202
54			0163	0167	0170	0174	0177	0181	0184	0187	0191	0194	0198	0201
57			0163	0166	0169	0173	0176	0180	0183	0186	0190	0193	0197	0200
10 0			0162	0165	0169	0172	0175	0179	0182	0186	0189	0192	0196	0199

TABLE V. LOG. A.

Logs. A, B, C, and D, for Lunars.

App. Alt. of ☽	REDUCED PARALLAX AND REFRACTION OF ☽														
	46'	47'	48'	49'	50'	51'	52'	53'	54'	55'	56'	57'	58'		
10° 0'	.0162	0165	0169	0172	0175	0179	0182	0186	0189	0192	0196	0199			
5	.0160	0164	0167	0171	0174	0177	0181	0184	0187	0191	0194	0197			
10	.0159	0162	0166	0169	0172	0176	0179	0182	0186	0189	0192	0196			
15	.0158	0161	0164	0168	0171	0174	0178	0181	0184	0187	0191	0194			
20	.0156	0160	0163	0166	0170	0173	0176	0179	0183	0186	0189	0192			
25	.0155	0158	0162	0165	0168	0171	0175	0178	0181	0184	0188	0191			
10 30	.0154	0157	0160	0164	0167	0170	0173	0177	0180	0183	0186	0189			
35	.0153	0156	0159	0162	0166	0169	0172	0175	0178	0181	0185	0188			
40	.0151	0155	0158	0161	0164	0167	0171	0174	0177	0180	0183	0186			
45	.0150	0153	0157	0160	0163	0166	0169	0172	0175	0179	0182	0185			
50	.0149	0152	0155	0158	0162	0165	0168	0171	0174	0177	0180	0183			
55	.0148	0151	0154	0157	0160	0163	0167	0170	0173	0176	0179	0182			
11 0	.0147	0150	0153	0156	0159	0162	0165	0168	0171	0174	0177	0181			
5	.0146	0149	0152	0155	0158	0161	0164	0167	0170	0173	0176	0179			
10		0148	0151	0154	0157	0160	0163	0166	0169	0172	0175	0178			
15		0146	0149	0152	0155	0158	0161	0164	0167	0170	0173	0176			
20		0145	0148	0151	0154	0157	0160	0163	0166	0169	0172	0175			
25		0144	0147	0150	0153	0156	0159	0162	0165	0168	0171	0174			
11 30		0143	0146	0149	0152	0155	0158	0161	0164	0167	0170	0172			
35		0142	0145	0148	0151	0154	0157	0160	0162	0165	0168	0171			
40		0141	0144	0147	0150	0153	0156	0158	0161	0164	0167	0170			
45		0140	0143	0146	0149	0151	0154	0157	0160	0163	0166	0169			
50		0139	0142	0145	0148	0150	0153	0156	0159	0162	0165	0167			
55		0138	0141	0144	0146	0149	0152	0155	0158	0161	0163	0166			
12 0		0137	0140	0143	0145	0148	0151	0154	0157	0159	0162	0165			
5		0136	0139	0142	0144	0147	0150	0153	0156	0158	0161	0164			
10		0135	0138	0141	0143	0146	0149	0152	0154	0157	0160	0163			
15		0134	0137	0140	0142	0145	0148	0151	0153	0156	0159	0162			
20		0133	0136	0139	0141	0144	0147	0150	0152	0155	0158	0160			
25		0132	0135	0138	0140	0143	0146	0148	0151	0154	0157	0159			
12 30		0131	0134	0137	0139	0142	0145	0147	0150	0153	0155	0158			
35		0130	0133	0136	0138	0141	0144	0146	0149	0152	0154	0157			
40		0129	0132	0135	0137	0140	0143	0145	0148	0151	0153	0156			
45		0129	0131	0134	0136	0139	0142	0144	0147	0150	0152	0155	0158		
50		0128	0130	0133	0136	0138	0141	0143	0146	0149	0151	0154	0156		
55		0127	0129	0132	0135	0137	0140	0142	0145	0148	0150	0153	0155		
13 0		0126	0129	0131	0134	0136	0139	0141	0144	0147	0149	0152	0154		
5		0125	0128	0130	0133	0135	0138	0141	0143	0146	0148	0151	0153		
10		0124	0127	0129	0132	0135	0137	0140	0142	0145	0147	0150	0152		
15		0123	0126	0129	0131	0134	0136	0139	0141	0144	0146	0149	0151		
20		0123	0125	0128	0130	0133	0135	0138	0140	0143	0145	0148	0150		
25		0122	0124	0127	0129	0132	0134	0137	0139	0142	0144	0147	0149		
13 30		0121	0124	0126	0129	0131	0133	0136	0138	0141	0143	0146	0148		
35		0120	0123	0125	0128	0130	0133	0135	0138	0140	0142	0145	0147		
40		0120	0122	0124	0127	0129	0132	0134	0137	0139	0142	0144	0146		
45			0121	0124	0126	0128	0131	0133	0136	0138	0141	0143	0145		
50			0120	0123	0125	0128	0130	0132	0135	0137	0140	0142	0145		
55			0120	0122	0124	0127	0129	0132	0134	0136	0139	0141	0144		
14 0			0119	0121	0124	0126	0128	0131	0133	0136	0138	0140	0143		
5			0118	0121	0123	0125	0128	0130	0132	0135	0137	0139	0142		
10			0117	0120	0122	0124	0127	0129	0132	0134	0136	0139	0141		
15			0117	0119	0121	0124	0126	0128	0131	0133	0135	0138	0140		
20			0116	0118	0121	0123	0125	0128	0130	0132	0135	0137	0139		
25			0115	0118	0120	0122	0124	0127	0129	0131	0134	0136	0138		
14 30			0114	0117	0119	0121	0124	0126	0128	0131	0133	0135	0137		
35			0114	0116	0118	0121	0123	0125	0128	0130	0132	0134	0137		
40			0113	0115	0118	0120	0122	0124	0127	0129	0131	0134	0136		
45			0112	0115	0117	0119	0121	0124	0126	0128	0130	0133	0135		
50			0112	0114	0116	0118	0121	0123	0125	0127	0130	0132	0134		
55			0111	0113	0116	0118	0120	0122	0124	0127	0129	0131	0133		
15 0			0110	0113	0115	0117	0119	0121	0124	0126	0128	0130	0133		

TABLE V. LOG. A.

Logs. A, B, C, and D, for Lunars.

App. Alt. of ☽		REDUCED PARALLAX AND REFRACTION OF ☽.													
	48'	49'	50'	51'	52'	53'	54'	55'	56'	57'	58'	59'			
15° 0'	.0110	0113	0115	0117	0119	0121	0124	0126	0128	0130	0133				
10	.0109	0111	0113	0116	0118	0120	0122	0124	0127	0129	0131				
20	.0108	0110	0112	0114	0116	0119	0121	0123	0125	0127	0129				
30	.0107	0109	0111	0113	0115	0117	0119	0121	0124	0126	0128				
40	.0105	0107	0110	0112	0114	0116	0118	0120	0122	0124	0126				
50	.0104	0106	0108	0110	0112	0115	0117	0119	0121	0123	0125				
16 0	.0103	0105	0107	0109	0111	0113	0115	0117	0119	0121	0124				
10	.0102	0104	0106	0108	0110	0112	0114	0116	0118	0120	0122				
20	.0101	0103	0105	0107	0109	0111	0113	0115	0117	0119	0121				
30	.0100	0102	0103	0105	0107	0109	0111	0113	0115	0117	0119				
40	.0098	0100	0102	0104	0106	0108	0110	0112	0114	0116	0118				
50	.0097	0099	0101	0103	0105	0107	0109	0111	0113	0115	0117				
17 0	.0096	0098	0100	0102	0104	0106	0108	0110	0112	0114	0116				
10	.0095	0097	0099	0101	0103	0105	0107	0109	0110	0112	0114				
20	.0094	0096	0098	0100	0102	0104	0106	0107	0109	0111	0113				
30		0095	0097	0099	0101	0103	0104	0106	0108	0110	0112				
40		0094	0096	0098	0100	0101	0103	0105	0107	0109	0111				
50		0093	0095	0097	0099	0100	0102	0104	0106	0108	0109				
18 0		0092	0094	0096	0098	0099	0101	0103	0105	0107	0108				
10		0091	0093	0095	0097	0098	0100	0102	0104	0105	0107	0109			
20		0090	0092	0094	0096	0097	0099	0101	0103	0104	0106	0108			
30		0089	0091	0093	0095	0096	0098	0100	0102	0103	0105	0107			
40		0088	0090	0092	0094	0095	0097	0099	0101	0102	0104	0106			
50		0088	0089	0091	0093	0094	0096	0098	0099	0101	0103	0105			
19 0		0087	0088	0090	0092	0093	0095	0097	0098	0100	0102	0104			
10		0086	0087	0089	0091	0092	0094	0096	0098	0099	0101	0103			
20		0085	0087	0088	0090	0092	0093	0095	0097	0098	0100	0102			
30		0084	0086	0087	0089	0091	0092	0094	0096	0097	0099	0101			
40		0083	0085	0087	0088	0090	0091	0093	0095	0096	0098	0100			
50		0082	0084	0086	0087	0089	0090	0092	0094	0095	0097	0099			
20 0		0082	0083	0085	0086	0088	0090	0091	0093	0094	0096	0098			
10		0081	0082	0084	0086	0087	0089	0090	0092	0093	0095	0097			
20		0080	0082	0083	0085	0086	0088	0089	0091	0093	0094	0096			
30		0079	0081	0082	0084	0086	0087	0089	0090	0092	0093	0095			
40		0079	0080	0082	0083	0085	0086	0088	0089	0091	0092	0094			
50		0078	0079	0081	0082	0084	0085	0087	0088	0090	0091	0093			
21 0		0077	0079	0080	0082	0083	0085	0086	0088	0089	0091	0092			
10		0076	0078	0079	0081	0082	0084	0085	0087	0088	0090	0091			
20		0076	0077	0079	0980	0082	0083	0085	0086	0087	0089	0090			
30		0075	0076	0078	0079	0081	0082	0084	0085	0087	0088	0090			
40		0074	0076	0077	0079	0080	0082	0083	0084	0086	0087	0089			
50		0074	0075	0076	0078	0079	0081	0082	0084	0085	0086	0088			
22 0		0073	0074	0076	0077	0079	0080	0081	0083	0084	0086	0087			
10		0072	0074	0075	0076	0078	0079	0081	0082	0083	0085	0086			
20		0072	0073	0074	0076	0077	0079	0080	0081	0083	0084	0086			
30		0071	0072	0074	0075	0076	0078	0079	0081	0082	0083	0085			
40		0070	0072	0073	0074	0076	0077	0079	0080	0081	0083	0084			
50		0070	0071	0072	0074	0075	0076	0078	0079	0081	0082	0083			
23 0		0069	0070	0072	0073	0074	0076	0077	0078	0080	0081	0082			
10		0068	0070	0071	0072	0074	0075	0076	0078	0079	0080	0082			
20		0068	0069	0070	0072	0073	0074	0076	0077	0078	0080	0081			
30		0067	0069	0070	0071	0072	0074	0075	0076	0078	0079	0080			
40		0067	0068	0069	0071	0072	0073	0074	0076	0077	0078	0080			
50		0066	0067	0069	0070	0071	0073	0074	0075	0076	0078	0079			
24 0			0067	0068	0069	0071	0072	0073	0074	0076	0077	0078			
10			0066	0067	0069	0070	0071	0073	0074	0075	0076	0078			
20			0066	0067	0068	0069	0071	0072	0073	0074	0076	0077			
30			0065	0066	0068	0069	0070	0071	0072	0074	0075	0076			
40			0065	0066	0067	0068	0069	0071	0072	0073	0074	0076			
50			0064	0065	0066	0068	0069	0070	0071	0072	0074	0075			
25 0			0063	0065	0066	0067	0068	0069	0071	0072	0073	0074			

TABLE V. LOG. A.

Logs. A, B, C, and D, for Lunars.

App. Alt. of ☽	\multicolumn{11}{c}{REDUCED PARALLAX AND REFRACTION OF ☽.}									
	50'	51'	52'	53'	54'	55'	56'	57'	58'	59'	60'
25° 0'	.0063	0065	0066	0067	0068	0069	0071	0072	0073	0074	
20	.0062	0064	0065	0066	0067	0068	0069	0071	0072	0073	
40	.0061	0062	0064	0065	0066	0067	0068	0069	0071	0072	
26 0	.0060	0061	0063	0064	0065	0066	0067	0068	0069	0071	
20	.0059	0060	0062	0063	0064	0065	0066	0067	0068	0069	
40	.0058	0059	0061	0062	0063	0064	0065	0066	0067	0068	
27 0	.0057	0058	0060	0061	0062	0063	0064	0065	0066	0067	
20	.0056	0057	0059	0060	0061	0062	0063	0064	0065	0066	
40	.0055	0057	0058	0059	0060	0061	0062	0063	0064	0065	
28 0	.0055	0056	0057	0058	0059	0060	0061	0062	0063	0064	
20	.0054	0055	0056	0057	0058	0059	0060	0061	0062	0063	
40	.0053	0054	0055	0056	0057	0058	0059	0060	0061	0062	
29 0	.0052	0053	0054	0055	0056	0057	0058	0059	0060	0061	
20	.0051	0052	0053	0054	0055	0056	0057	0058	0059	0060	
40	.0050	0051	0052	0053	0054	0055	0056	0057	0058	0059	
30 0	.0050	0051	0051	0052	0053	0054	0055	0056	0057	0058	
20	.0049	0050	0051	0052	0052	0053	0054	0055	0056	0057	
40	.0048	0049	0050	0051	0052	0053	0053	0054	0055	0056	
31 0	.0047	0048	0049	0050	0051	0052	0053	0053	0054	0055	
20	.0047	0047	0048	0049	0050	0051	0052	0053	0054	0054	0055
40	.0046	0047	0048	0048	0049	0050	0051	0052	0053	0054	0054
32 0	.0045	0046	0047	0048	0048	0049	0050	0051	0052	0053	0054
20	.0044	0045	0046	0047	0048	0049	0049	0050	0051	0052	0053
40	.0044	0045	0045	0046	0047	0048	0049	0049	0050	0051	0052
33 0	.0043	0044	0045	0045	0046	0047	0048	0049	0049	0050	0051
20	.0042	0043	0044	0045	0046	0046	0047	0048	0049	0050	0050
40	.0042	0043	0043	0044	0045	0046	0046	0047	0048	0049	0050
34 0	.0041	0042	0043	0043	0044	0045	0046	0046	0047	0048	0049
20	.0040	0041	0042	0043	0043	0044	0045	0046	0047	0047	0048
40	.0040	0041	0041	0042	0043	0044	0044	0045	0046	0047	0047
35 0	.0039	0040	0041	0041	0042	0043	0044	0044	0045	0046	0047
20	.0039	0039	0040	0041	0042	0042	0043	0044	0044	0045	0046
40	.0038	0039	0039	0040	0041	0042	0042	0043	0044	0044	0045
36 0	.0037	0038	0039	0040	0040	0041	0042	0042	0043	0044	0044
20	.0037	0038	0038	0039	0040	0040	0041	0042	0042	0043	0044
40	.0036	0037	0038	0038	0039	0040	0040	0041	0042	0042	0043
37 0	.0036	0036	0037	0038	0038	0039	0040	0040	0041	0042	0042
20	.0035	0036	0037	0037	0038	0039	0039	0040	0040	0041	0042
40	.0035	0035	0036	0037	0037	0038	0039	0039	0040	0040	0041
38 0	.0034	0035	0035	0036	0037	0037	0038	0039	0039	0040	0040
20	.0034	0034	0035	0036	0036	0037	0037	0038	0039	0039	0040
40	.0033	0034	0034	0035	0036	0036	0037	0037	0038	0039	0039
39 0		0033	0034	0034	0035	0036	0036	0037	0037	0038	0039
20		0033	0033	0034	0035	0035	0036	0036	0037	0037	0038
40		0032	0033	0033	0034	0035	0035	0036	0036	0037	0037
40 0		0032	0032	0033	0033	0034	0035	0035	0036	0036	0037
20		0031	0032	0032	0033	0034	0034	0035	0035	0036	0036
40		0031	0031	0032	0032	0033	0034	0034	0035	0035	0036
41 0		0030	0031	0031	0032	0032	0033	0034	0034	0035	0035
20		0030	0030	0031	0031	0032	0033	0033	0034	0034	0035
40		0029	0030	0030	0031	0032	0032	0033	0033	0034	0034
42 0		0029	0029	0030	0031	0031	0032	0032	0033	0033	0034
20		0029	0029	0030	0030	0031	0031	0032	0032	0033	0033
40		0028	0029	0029	0030	0030	0031	0031	0032	0032	0033
43 0		0028	0028	0029	0029	0030	0030	0031	0031	0032	0032
20		0027	0028	0028	0029	0029	0030	0030	0031	0031	0032
40		0027	0027	0028	0028	0029	0029	0030	0030	0031	0031
44 0		0026	0027	0027	0028	0028	0029	0029	0030	0030	0031
20		0026	0026	0027	0027	0028	0028	0029	0029	0030	0030
40		0026	0026	0026	0027	0027	0028	0028	0029	0029	0030
45 0		0025	0026	0026	0027	0027	0027	0028	0028	0029	0029

TABLE V. LOG. A.

Logs. A, B, C, and D. for Lunars.

App. Alt. of ☽	REDUCED PARALLAX AND REFRACTION OF ☽.													
	51'	52'	53'	54'	55'	56'	57'	58'	59'	60'				
45° 0'	.0025	0026	0026	0027	0027	0027	0028	0028	0029	0029				
30	.0025	0025	0025	0026	0026	0027	0027	0028	0028	0028				
46 0	.0024	0024	0025	0025	0026	0026	0027	0027	0027	0028				
30	.0023	0024	0024	0025	0025	0026	0026	0026	0027	0027				
47 0	.0023	0023	0024	0024	0025	0025	0025	0026	0026	0026				
30	.0022	0023	0023	0024	0024	0024	0025	0025	0025	0026				
48 0	.0022	0022	0023	0023	0023	0024	0024	0024	0025	0025				
30	.0021	0022	0022	0022	0023	0023	0024	0024	0024	0025				
49 0	.0021	0021	0022	0022	0022	0023	0023	0023	0024	0024				
30	.0020	0021	0021	0021	0022	0022	0022	0023	0023	0023				
50 0	.0020	0020	0020	0021	0021	0022	0022	0022	0023	0023				
30	.0019	0020	0020	0020	0021	0021	0021	0022	0022	0022				
51 0	.0019	0019	0020	0020	0020	0020	0021	0021	0021	0022				
30	.0018	0019	0019	0019	0020	0020	0020	0021	0021	0021				
52 0	.0018	0018	0019	0019	0019	0019	0020	0020	0020	0021				
30	.0018	0018	0018	0018	0019	0019	0019	0020	0020	0020				
53 0	.0017	0017	0018	0018	0018	0018	0019	0019	0019	0020				
30	.0017	0017	0017	0017	0018	0018	0018	0019	0019	0019				
54 0	.0016	0016	0017	0017	0017	0018	0018	0018	0018	0019				
30	.0016	0016	0016	0017	0017	0017	0017	0018	0018	0018				
55 0	.0015	0016	0016	0016	0016	0017	0017	0017	0017	0018				
30	.0015	0015	0015	0016	0016	0016	0016	0017	0017	0017				
56 0	.0015	0015	0015	0015	0016	0016	0016	0016	0017	0017				
30	.0014	0014	0015	0015	0015	0015	0016	0016	0016	0016				
57 0	.0014	0014	0014	0015	0015	0015	0015	0015	0016	0016				
30	.0014	0014	0014	0014	0014	0015	0015	0015	0015	0015				
58 0	.0013	0013	0014	0014	0014	0014	0014	0015	0015	0015				
30	.0013	0013	0013	0013	0014	0014	0014	0014	0014	0015				
59 0	.0012	0013	0013	0013	0013	0013	0014	0014	0014	0014				
30	.0012	0012	0012	0013	0013	0013	0013	0013	0014	0014				
60	.0012	0012	0012	0012	0013	0013	0013	0013	0013	0013				
61	.0011	0011	0011	0012	0012	0012	0012	0012	0012	0013				
62	.0011	0011	0011	0011	0011	0011	0011	0012	0012	0012				
63	.0010	0010	0010	0010	0011	0011	0011	0011	0011	0011				
64	.0009	0010	0010	0010	0010	0010	0010	0010	0010	0011				
65	.0009	0009	0009	0009	0009	0009	0010	0010	0010	0010				
66	.0008	0008	0009	0009	0009	0009	0009	0009	0009	0009				
67	.0008	0008	0008	0008	0008	0008	0008	0009	0009	0009				
68	.0007	0007	0008	0008	0008	0008	0008	0008	0008	0008				
69	.0007	0007	0007	0007	0007	0007	0007	0008	0008	0008				
70	.0007	0007	0007	0007	0007	0007	0007	0007	0007	0007				
71	.0006	0006	0006	0006	0006	0006	0007	0007	0007	0007				
72	.0006	0006	0006	0006	0006	0006	0006	0006	0006	0006				
73	.0005	0005	0006	0006	0006	0006	0006	0006	0006	0006				
74	.0005	0005	0005	0005	0005	0005	0005	0005	0005	0006				
75	.0005	0005	0005	0005	0005	0005	0005	0005	0005	0005				
76	.0004	0005	0005	0005	0005	0005	0005	0005	0005	0005				
77	.0004	0004	0004	0004	0004	0004	0004	0004	0004	0004				
78	.0004	0004	0004	0004	0004	0004	0004	0004	0004	0004				
79	.0004	0004	0004	0004	0004	0004	0004	0004	0004	0004				
80	.0004	0004	0004	0004	0004	0004	0004	0004	0004	0004				
81	.0003	0003	0003	0003	0003	0003	0003	0003	0003	0003				
82	.0003	0003	0003	0003	0003	0003	0003	0003	0003	0003				
83	.0003	0003	0003	0003	0003	0003	0003	0003	0003	0003				
84	.0003	0003	0003	0003	0003	0003	0003	0003	0003	0003				
85	.0003	0003	0003	0003	0003	0003	0003	0003	0003	0003				
86	.0003	0003	0003	0003	0003	0003	0003	0003	0003	0003				
87	.0003	0003	0003	0003	0003	0003	0003	0003	0003	0003				
88	0003	0003	0003	0003	0003	0003	0003	0003	0003	0003				
89	.0003	0003	0003	0003	0003	0003	0003	0003	0003	0003				
90	.0003	0003	0003	0003	0003	0003	0003	0003	0003	0003				

TABLE V. LOG. B.

Logs. A, B, C, and D, for Lunars.

REDUCED REFRACTION AND PARALLAX OF ⊙ OR ✳.

App. Alt. of ⊙ or ✳	0' 0"	0' 30"	1' 0"	1' 30"	2' 0"	2' 30"	3' 0"	3' 30"	4' 0"	4' 30"	5' 0"	5' 30"
5° 0'												
10												
20												
30												
40												
50												
6 0												
20												9.9970
40												9.9972
7 0											9.9976	9.9974
20											9.9977	9.9975
40										9.9981	9.9978	9.9976
8 0										9.9982	9.9979	9.9977
20										9.9982	9.9980	9.9978
40										9.9983	9.9981	9.9979
9 0								9.9986		9.9984	9.9982	9.9980
20								9.9986		9.9985	9.9983	9.9981
40								9.9987		9.9985	9.9983	9.9982
10 0							9.9989	9.9988		9.9986	9.9984	9.9982
11							9.9992	9.9991	9.9989	9.9987	9.9986	9.9984
12							9.9993	9.9992	9.9990	9.9989	9.9987	9.9986
13						9.9995	9.9994	9.9992	9.9991	9.9990	9.9989	9.9987
14						9.9995	9.9994	9.9993	9.9992	9.9991	9.9990	
15					9.9997	9.9996	9.9995	9.9994	9.9993	9.9992	9.9991	
16					9.9997	9.9996	9.9995	9.9994	9.9993	9.9993		
18				9.9990	9.9998	9.9997	9.9996	9.9995	9.9995			
20			0.0000	9.9999	9.9998	9.9998	9.9997	9.9996	9.9996			
25			0.0000	0.0000	9.9999	9.9999	9.9998	9.9998				
30		0.0001	0.0001	0.0000	0.0000	0.0000	9.9999					
50	0.0001	0.0001	0.0001	0.0001	0.0001	0.0001						
90	0.0001	0.0002	0.0002	0.0002								

LOG. C.

REDUCED REFRACTION AND PARALLAX OF ⊙ OR ✳.

App. Alt. of ⊙ or ✳	0' 0"	0' 30"	1' 0"	1' 30"	2' 0"	2' 30"	3' 0"	3' 30"	4' 0"	4' 30"	5' 0"	5' 30"
5° 0'												
20												
40												
6 0												
20												9.9969
40												9.9970
7 0											9.9974	9.9972
8										9.9980	9.9978	9.9975
9								9.9984		9.9982	9.9980	9.9978
10							9.9988	9.9986		9.9984	9.9982	9.9981
11							9.9990	9.9989	9.9987	9.9986	9.9984	9.9982
12							9.9991	9.9990	9.9988	9.9987	9.9985	9.9984
13							9.9992	9.9991	9.9989	9.9987	9.9985	
14						9.9994	9.9993	9.9991	9.9990	9.9989	9.9988	
15					9.9995	9.9994	9.9993	9.9992	9.9991	9.9990	9.9989	
16					9.9996	9.9995	9.9994	9.9993	9.9992	9.9990		
17					9.9996	9.9995	9.9994	9.9993	9.9992	9.9991		
18				9.9997	9.9996	9.9995	9.9994	9.9994	9.9993			
20			9.9998	9.9998	9.9997	9.9996	9.9995	9.9994	9.9993			
25			9.9999	9.9998	9.9998	9.9997	9.9996	9.9996				
30		0.0000	9.9999	9.9999	9.9999	9.9998	9.9997					
40		0.0000	9.9999	9.9999	9.9999	9.9990						
50	0.0000	0.0000	0.0000	9.9999	9.9999							
90	0.0000	0.0000	0.0000	0.0000								

TABLE V. LOG. B.

Logs. A, B, C, and D, for Lunars.

REDUCED REFRACTION AND PARALLAX OF ⊙ OR ☀.

App. Alt. of ⊙ or ☀	6' 0"	6' 30"	7' 0"	7' 30"	8' 0"	8' 30"	9' 0"	9' 30"	10' 0"	10' 30"	11' 0"	11' 30"
5° 0'			9.9951	9.9947	9.9944	9.9940	9.9937	9.9933	9.9929	9.9926	9.9922	9.9919
10			9.9953	9.9949	9.9946	9.9942	9.9939	9.9935	9.9932	9.9928	9.9925	9.9921
20			9.9954	9.9951	9.9948	9.9944	9.9941	9.9937	9.9934	9.9931	9.9927	9.9924
30		9.9959	9.9956	9.9952	9.9949	9.9946	9.9943	9.9939	9.9936	9.9933	9.9929	
40		9.9960	9.9957	9.9954	9.9951	9.9948	9.9944	9.9941	9.9938	9.9935	9.9932	
50	9.9965	9.9962	9.9958	9.9955	9.9952	9.9949	9.9946	9.9943	9.9940	9.9937		
6 0	9.9966	9.9963	9.9960	9.9957	9.9954	9.9951	9.9948	9.9945	9.9942	9.9939		
20	9.9968	9.9965	9.9962	9.9959	9.9956	9.9954	9.9951	9.9948	9.9945			
40	9.9969	9.9967	9.9964	9.9961	9.9959	9.9956	9.9953	9.9951	9.9948			
7 0	9.9971	9.9968	9.9966	9.9963	9.9961	9.9958	9.9956	9.9953				
20	9.9972	9.9970	9.9968	9.9965	9.9963	9.9960	9.9958					
40	9.9974	9.9971	9.9969	9.9967	9.9965	9.9962						
8 0	9.9975	9.9973	9.9971	9.9968	9.9966	9.9964						
20	9.9976	9.9974	9.9972	9.9970	9.9968							
40	9.9977	9.9975	9.9973	9.9971								
9 0	9.9978	9.9976	9.9974	9.9972								
20	9.9979	9.9977	9.9975									
40	9.9980	9.9978	9.9976									
10	9.9981	9.9979	9.9977									
11	9.9983	9.9981										
12	9.9985											
13												
14												
15												
16												
18												
20												
25												
30												
50												
90												

LOG. C.

REDUCED REFRACTION AND PARALLAX OF ⊙ OR ☀.

App. Alt. of ⊙ or ☀	6' 0"	6' 30"	7' 0"	7' 30"	8' 0"	8' 30"	9' 0"	9' 30"	10' 0"	10' 30"	11' 0"	11' 30"
5° 0'			9.9949	9.9946	9.9942	9.9938	9.9935	9.9931	9.9927	9.9924	9.9920	9.9916
20		9.9956	9.9953	9.9949	9.9946	9.9942	9.9939	9.9936	9.9932	9.9929	9.9925	9.9922
40	9.9962	9.9959	9.9955	9.9952	9.9949	9.9946	9.9943	9.9939	9.9936	9.9933	9.9930	
6 0	9.9964	0.9961	9.9958	9.9955	9.9952	9.9949	9.9946	9.9943	9.9940	9.9937		
20	9.9966	9.9963	9.9960	9.9957	9.9955	9.9952	9.9949	9.9946	9.9943			
40	9.9968	9.9965	9.9962	9.9960	9.9957	9.9954	9.9951	9.9949	9.9946			
7	9.9969	9.9967	9.9964	9.9962	9.9959	9.9956	9.9954	9.9951				
8	9.9973	9.9971	9.9969	9.9966	9.9964	9.9962	9.9960					
9	9.9976	9.9974	9.9972	9.9970	9.9968							
10	9.9979	9.9977	9.9975									
11	9.9981	9.9979										
12	9.9983											
13												
14												
15												
16												
17												
18												
20												
25												
30												
40												
50												
90												

TABLE V. LOG. D.

Logs. A, B, C, and D, for Lunars.

App. Alt. of ☽	REDUCED PARALLAX AND REFRACTION OF ☽ 41'	42'	43'	44'	45'	46'	47'	48'	49'	50'	51'	52'	53'	54'	55'
5° 0'	.0283	0290	0296	0303	0310	0316	0323	0329	0336	0343	0349	0356	0362	0369	
3	.0280	0287	0293	0300	0307	0313	0320	0326	0333	0339	0346	0352	0359	0365	
6	.0277	0284	0291	0297	0304	0310	0317	0323	0330	0336	0342	0349	0355	0362	
9	.0275	0281	0288	0294	0301	0307	0313	0320	0326	0333	0339	0345	0352	0358	
12	.0272	0279	0285	0291	0298	0304	0310	0317	0323	0330	0336	0342	0349	0355	
5 15	.0270	0276	0282	0289	0295	0301	0308	0314	0320	0326	0333	0339	0345	0351	
18	.0267	0273	0280	0286	0292	0298	0305	0311	0317	0323	0330	0336	0342	0348	
21	.0264	0271	0277	0283	0289	0296	0302	0308	0314	0320	0327	0333	0339	0345	
24	.0262	0268	0274	0281	0287	0293	0299	0305	0311	0317	0324	0330	0336	0342	
27	.0260	0266	0272	0278	0284	0290	0296	0302	0308	0314	0321	0327	0333	0339	
5 30	.0257	0263	0269	0275	0282	0288	0294	0300	0306	0312	0318	0324	0330	0336	
33	.0255	0261	0267	0273	0279	0285	0291	0297	0303	0309	0315	0321	0327	0333	
36	.0253	0259	0265	0271	0276	0282	0288	0294	0300	0306	0312	0318	0324	0330	
39		0256	0262	0268	0274	0280	0286	0292	0298	0303	0309	0315	0321	0327	
42		0254	0260	0266	0272	0277	0283	0289	0295	0301	0306	0312	0318	0324	
5 45		0252	0258	0263	0269	0275	0281	0287	0292	0298	0304	0310	0315	0321	
48		0250	0255	0261	0267	0273	0278	0284	0290	0295	0301	0307	0313	0318	
51		0247	0253	0259	0265	0270	0276	0282	0287	0293	0299	0304	0310	0316	
54		0245	0251	0257	0262	0268	0274	0279	0285	0290	0296	0302	0307	0313	
57		0243	0249	0254	0260	0266	0271	0277	0282	0288	0294	0299	0305	0310	
6 0		0241	0247	0252	0258	0263	0269	0275	0280	0286	0291	0297	0302	0308	
3		0239	0245	0250	0256	0261	0267	0272	0278	0283	0289	0294	0300	0305	
6		0237	0243	0248	0254	0259	0265	0270	0275	0281	0286	0292	0297	0302	
9		0235	0241	0246	0252	0257	0262	0268	0273	0279	0284	0289	0295	0300	
12		0233	0239	0244	0249	0255	0260	0266	0271	0276	0282	0287	0292	0298	
6 15		0231	0237	0242	0247	0253	0258	0263	0269	0274	0279	0285	0290	0295	
18		0230	0235	0240	0245	0251	0256	0261	0267	0272	0277	0282	0288	0293	
21		0228	0233	0238	0243	0249	0254	0259	0264	0270	0275	0280	0285	0290	
24		0226	0231	0236	0242	0247	0252	0257	0262	0267	0273	0278	0283	0288	
27			0229	0234	0240	0245	0250	0255	0260	0265	0271	0276	0281	0286	0291
6 30			0227	0233	0238	0243	0248	0253	0258	0263	0268	0274	0279	0284	0289
33			0226	0231	0236	0241	0246	0251	0256	0261	0266	0271	0276	0281	0287
36			0224	0229	0234	0239	0244	0249	0254	0259	0264	0269	0274	0279	0284
39			0222	0227	0232	0237	0242	0247	0252	0257	0262	0267	0272	0277	0282
42			0220	0225	0230	0235	0240	0245	0250	0255	0260	0265	0270	0275	0280
6 45			0219	0224	0229	0234	0239	0244	0248	0253	0258	0263	0268	0273	0278
48			0217	0222	0227	0232	0237	0242	0247	0251	0256	0261	0266	0271	0276
51			0216	0220	0225	0230	0235	0240	0245	0250	0254	0259	0264	0269	0274
54			0214	0219	0224	0228	0233	0238	0243	0248	0253	0257	0262	0267	1272
57			0212	0217	0222	0227	0232	0236	0241	0246	0251	0255	0260	0265	0270
7 0			0211	0216	0220	0225	0230	0235	0239	0244	0249	0254	0258	0263	0268
3			0209	0214	0219	0223	0228	0233	0238	0242	0247	0252	0256	0261	0266
6			0208	0212	0217	0222	0227	0231	0236	0241	0245	0250	0255	0259	0264
9				0211	0216	0220	0225	0230	0234	0239	0243	0248	0253	0257	0262
12				0209	0214	0219	0223	0228	0232	0237	0242	0246	0251	0255	0260
7 15				0208	0212	0217	0222	0226	0231	0235	0240	0245	0249	0254	0258
18				0206	0211	0216	0220	0225	0229	0234	0238	0243	0247	0252	0256
21				0205	0209	0214	0219	0223	0228	0232	0237	0241	0246	0250	0255
24				0204	0208	0213	0217	0222	0226	0230	0235	0239	0244	0248	0253
27				0202	0207	0211	0216	0220	0224	0229	0233	0238	0242	0247	0251
7 30				0201	0205	0210	0214	0218	0223	0227	0232	0236	0241	0245	0249
33				0199	0204	0208	0213	0217	0221	0226	0230	0234	0239	0243	0248
36				0198	0202	0207	0211	0215	0220	0224	0229	0233	0237	0242	0246
39				0197	0201	0205	0210	0214	0218	0223	0227	0231	0236	0240	0244
42				0195	0200	0204	0208	0213	0217	0221	0225	0230	0234	0238	0243
7 45				0194	0198	0203	0207	0211	0215	0220	0224	0228	0232	0237	0241
48				0193	0197	0201	0205	0210	0214	0218	0222	0227	0231	0235	0239
51				0191	0196	0200	0204	0208	0213	0217	0221	0225	0229	0234	0238
54				0190	0194	0198	0203	0207	0211	0215	0219	0224	0228	0232	0236
57				0189	0193	0197	0201	0206	0210	0214	0218	0222	0226	0230	0235
8 0				0188	0192	0196	0200	0204	0208	0212	0217	0221	0225	0229	0233

TABLE V. LOG. D.

Logs. A, B, C, and D, for Lunars.

App. Alt. of ☽	REDUCED PARALLAX AND REFRACTION OF ☽.													
	45′	46′	47′	48′	49′	50′	51′	52′	53′	54′	55′	56′	57′	58
8 0	.0192	0196	0200	0204	0208	0212	0217	0221	0225	0229	0233	0237		
5	.0190	0194	0198	0202	0206	0210	0214	0218	0222	0227	0231	0235		
10	.0188	0192	0196	0200	0204	0208	0212	0216	0220	0224	0228	0232		
15	.0186	0190	0194	0198	0202	0206	0210	0214	0218	0222	0226	0230		
20	.0184	0188	0192	0196	0200	0204	0207	0211	0215	0219	0223	0227		
25	.0182	0186	0190	0194	0197	0201	0205	0209	0213	0217	0221	0225		
8 30	.0180	0184	0188	0192	0195	0199	0203	0207	0211	0215	0219	0223		
35	.0178	0182	0186	0190	0193	0197	0201	0205	0209	0213	0216	0220		
40	.0176	0180	0184	0188	0191	0195	0199	0203	0207	0210	0214	0218		
45	.0174	0178	0182	0186	0189	0193	0197	0201	0205	0208	0212	0216		
50	.0173	0176	0180	0184	0188	0191	0195	0199	0202	0206	0210	0214		
55	.0171	0175	0178	0182	0186	0189	0193	0197	0200	0204	0208	0212		
9 0	.0169	0173	0177	0180	0184	0188	0191	0195	0198	0202	0206	0209		
5	.0167	0171	0175	0178	0182	0186	0189	0193	0197	0200	0204	0207		
10	.0166	0169	0173	0177	0180	0184	0187	0191	0195	0198	0202	0205		
15	.0164	0168	0171	0175	0179	0182	0186	0189	0193	0196	0200	0203		
20	.0163	0166	0170	0173	0177	0180	0184	0187	0191	0194	0198	0201		
25	.0161	0165	0168	0172	0175	0179	0182	0186	0189	0193	0196	0199		
9 30		0163	0166	0170	0173	0177	0180	0184	0187	0191	0194	0198		
35		0161	0165	0168	0172	0175	0179	0182	0185	0189	0192	0196		
40		0160	0163	0167	0170	0174	0177	0180	0184	0187	0191	0194		
45		0158	0162	0165	0169	0172	0175	0179	0182	0185	0189	0192	0195	
50		0157	0160	0164	0167	0170	0174	0177	0180	0184	0187	0190	0194	
55		0156	0159	0162	0165	0169	0172	0175	0179	0182	0185	0189	0192	
10 0		0154	0157	0161	0164	0167	0171	0174	0177	0180	0184	0187	0190	
5		0153	0156	0159	0162	0166	0169	0172	0175	0179	0182	0185	0188	
10		0151	0155	0158	0161	0164	0167	0171	0174	0177	0180	0183	0187	
15		0150	0153	0156	0160	0163	0166	0169	0172	0175	0179	0182	0185	
20		0149	0152	0155	0158	0161	0164	0168	0171	0174	0177	0180	0183	
25		0147	0150	0154	0157	0160	0163	0166	0169	0172	0175	0179	0182	
10 30		0146	0149	0152	0155	0158	0162	0165	0168	0171	0174	0177	0180	
35		0145	0148	0151	0154	0157	0160	0163	0166	0169	0172	0175	0179	
40		0143	0147	0150	0153	0156	0159	0162	0165	0168	0171	0174	0177	
45		0142	0145	0148	0151	0154	0157	0160	0163	0166	0169	0172	0175	
50		0141	0144	0147	0150	0153	0156	0159	0162	0165	0168	0171	0174	
55		0140	0143	0146	0149	0152	0155	0158	0161	0164	0167	0170	0172	
11 0		0139	0142	0145	0147	0150	0153	0156	0159	0162	0165	0168	0171	
5		0137	0140	0143	0146	0149	0152	0155	0158	0161	0164	0167	0170	
10			0139	0142	0145	0148	0151	0154	0157	0159	0162	0165	0168	
15			0138	0141	0144	0147	0150	0152	0155	0158	0161	0164	0167	
20			0137	0140	0143	0145	0148	0151	0154	0157	0160	0163	0165	
25			0136	0139	0141	0144	0147	0150	0153	0156	0158	0161	0164	
11 30			0135	0137	0140	0143	0146	0149	0151	0154	0157	0160	0163	
35			0133	0136	0139	0142	0145	0147	0150	0153	0156	0159	0161	
40			0132	0135	0138	0141	0143	0146	0149	0152	0154	0157	0160	
45			0131	0134	0137	0140	0142	0145	0148	0150	0153	0156	0159	
50			0130	0133	0136	0138	0141	0144	0147	0149	0152	0155	0157	
55			0129	0132	0135	0137	0140	0143	0145	0148	0151	0153	0156	
12 0			0128	0131	0134	0136	0139	0142	0144	0147	0150	0152	0155	
5			0127	0130	0132	0135	0138	0140	0143	0146	0148	0151	0154	
10			0126	0129	0131	0134	0137	0139	0142	0145	0147	0150	0152	
15			0125	0128	0130	0133	0136	0138	0141	0143	0146	0149	0151	
20			0124	0127	0129	0132	0135	0137	0140	0142	0145	0147	0150	
25			0123	0126	0128	0131	0133	0136	0139	0141	0144	0146	0149	
12 30			0122	0125	0127	0130	0132	0135	0138	0140	0143	0145	0148	
35			0121	0124	0126	0129	0131	0134	0136	0139	0141	0144	0147	
40			0120	0123	0125	0128	0130	0133	0135	0138	0140	0143	0145	
45			0119	0122	0124	0127	0129	0132	0134	0137	0139	0142	0144	0147
50			0118	0121	0123	0126	0128	0131	0133	0136	0138	0141	0143	0146
55			0118	0120	0123	0125	0127	0130	0132	0135	0137	0140	0142	0145
13 0			0117	0119	0122	0124	0126	0129	0131	0134	0136	0139	0141	0143

TABLE V. LOG. D.

Logs. A. B. C. and D, for Lunars.

App. Alt. of ☽	47'	48'	49'	50'	51'	52'	53'	54'	55'	56'	57'	58'	59'		
13° 0'	.0117	0119	0122	0124	0126	0129	0131	0134	0136	0139	0141	0143			
10	.0115	0117	0120	0122	0125	0127	0129	0132	0134	0137	0139	0141			
20	.0113	0116	0118	0120	0123	0125	0127	0130	0132	0134	0137	0139			
30	.0112	0114	0116	0119	0121	0123	0125	0128	0130	0132	0135	0137			
40		0112	0114	0117	0119	0121	0124	0126	0128	0131	0133	0135			
50		0111	0113	0115	0117	0120	0122	0124	0126	0129	0131	0133			
14 0		0109	0111	0113	0116	0118	0120	0122	0125	0127	0129	0131			
10		0107	0110	0112	0114	0116	0118	0121	0123	0125	0127	0129			
20		0106	0108	0110	0112	0114	0117	0119	0121	0123	0125	0127			
30		0104	0106	0109	0111	0113	0115	0117	0119	0121	0123	0126			
40		0103	0105	0107	0109	0111	0113	0115	0118	0120	0122	0124			
50		0101	0103	0106	0108	0110	0112	0114	0116	0118	0120	0122			
15 0		0100	0102	0104	0106	0108	0110	0112	0114	0116	0118	0120			
10		0099	0101	0103	0105	0107	0109	0111	0113	0115	0117	0119			
20		0097	0099	0101	0103	0105	0107	0109	0111	0113	0115	0117			
30		0096	0098	0100	0102	0104	0106	0108	0110	0112	0113	0115			
40		0094	0096	0098	0100	0102	0104	0106	0108	0110	0112	0114			
50		0093	0095	0097	0099	0101	0103	0105	0107	0108	0110	0112			
16 0		0092	0094	0096	0098	0099	0101	0103	0105	0107	0109	0111			
10		0091	0093	0094	0096	0098	0100	0102	0104	0106	0107	0109			
20		0089	0091	0093	0095	0097	0099	0100	0102	0104	0106	0108			
30		0088	0090	0092	0094	0096	0097	0099	0101	0103	0105	0106			
40		0087	0089	0091	0092	0094	0096	0098	0100	0101	0103	0105			
50		0086	0088	0089	0091	0093	0095	0096	0098	0100	0102	0104			
17 0		0085	0087	0088	0090	0092	0093	0095	0097	0099	0100	0102			
10		0084	0085	0087	0089	0091	0092	0094	0096	0097	0099	0101			
20		0083	0084	0086	0088	0089	0091	0093	0094	0096	0098	0099			
30			0083	0085	0086	0088	0090	0091	0093	0095	0096	0098			
40			0082	0084	0085	0087	0089	0090	0092	0094	0095	0097			
50			0081	0083	0084	0086	0087	0089	0091	0092	0094	0096			
18 0			0080	0082	0083	0085	0086	0088	0090	0091	0093	0094			
20			0078	0079	0081	0083	0084	0086	0087	0089	0090	0092	0093		
40			0076	0077	0079	0080	0082	0083	0085	0087	0088	0090	0091		
19 0			0074	0075	0077	0078	0080	0081	0083	0084	0086	0087	0089		
20			0072	0073	0075	0076	0078	0079	0081	0082	0084	0085	0086		
40			0070	0072	0073	0074	0076	0077	0079	0080	0081	0083	0084		
20 0			0068	0070	0071	0073	0074	0075	0077	0078	0079	0081	0082		
20			0067	0068	0069	0071	0072	0073	0075	0076	0077	0079	0080		
40			0065	0066	0068	0069	0070	0072	0073	0074	0075	0077	0078		
21 0			0063	0065	0066	0067	0068	0070	0071	0072	0074	0075	0076		
20			0062	0063	0064	0065	0067	0068	0069	0070	0072	0073	0074		
40			0060	0061	0063	0064	0065	0066	0067	0069	0070	0071	0072		
22 0			0059	0060	0061	0062	0063	0065	0066	0067	0068	0069	0070		
20			0057	0058	0059	0061	0062	0063	0064	0065	0066	0068	0069		
40			0056	0057	0058	0059	0060	0061	0062	0063	0065	0066	0067		
23 0			0054	0055	0057	0058	0059	0060	0061	0062	0063	0064	0065		
20			0053	0054	0055	0056	0057	0058	0059	0060	0061	0063	0064		
40			0052	0053	0054	0055	0056	0057	0058	0059	0060	0061	0062		
24 0			0050	0051	0052	0053	0054	0055	0056	0057	0058	0059	0060		
20				0050	0051	0052	0053	0054	0055	0056	0057	0058	0059		
40				0049	0050	0051	0052	0053	0053	0054	0055	0056	0057		
25 0				0047	0048	0049	0049	0050	0051	0052	0053	0054	0055		
20				0046	0047	0048	0049	0050	0051	0052	0053	0053	0054		
40				0045	0046	0047	0048	0049	0049	0050	0051	0052	0053		
26 0				0044	0045	0046	0046	0047	0048	0049	0050	0051	0052		
20				0043	0043	0044	0045	0046	0047	0048	0048	0049	0050		
40				0041	0042	0043	0044	0045	0046	0046	0047	0048	0049		
27 0				0040	0041	0042	0043	0044	0044	0045	0046	0047	0047		
20				0039	0040	0041	0042	0042	0043	0044	0045	0046	0046		
40				0038	0039	0040	0040	0041	0042	0043	0043	0044	0045		
28 0				0037	0039	0039	0039	0040	0041	0042	0042	0043	0044		

34

TABLE V. LOG. D.

Logs. A, B, C, and D, for Lunars.

Apparent Altitude of ☽	\multicolumn{11}{c	}{REDUCED PARALLAX AND REFRACTION OF ☽.}
	50'	51'	52'	53'	54'	55'	56'	57'	58'	59'	60'
28° 0'	0.0037	0.0038	0.0039	0.0039	0.0040	0.0041	0.0042	0.0042	0.0043	0.0044	
30	0.0036	0.0036	0.0037	0.0038	0.0038	0.0039	0.0040	0.0040	0.0041	0.0042	
29 0	0.0034	0.0035	0.0035	0.0036	0.0037	0.0037	0.0038	0.0039	0.0039	0.0040	
30	0.0033	0.0033	0.0034	0.0035	0.0035	0.0036	0.0036	0.0037	0.0038	0.0038	
30 0	0.0031	0.0032	0.0032	0.0033	0.0034	0.0034	0.0035	0.0035	0.0036	0.0037	
30	0.0030	0.0030	0.0031	0.0031	0.0032	0.0033	0.0033	0.0034	0.0034	0.0035	
31 0	0.0028	0.0029	0.0029	0.0030	0.0031	0.0031	0.0032	0.0032	0.0033	0.0033	
30	0.0027	0.0028	0.0028	0.0029	0.0029	0.0030	0.0030	0.0031	0.0031	0.0032	0.0032
32 0	0.0026	0.0026	0.0027	0.0027	0.0028	0.0028	0.0029	0.0029	0.0030	0.0030	0.0031
30	0.0024	0.0025	0.0025	0.0026	0.0026	0.0027	0.0027	0.0028	0.0028	0.0029	0.0029
33 0	0.0023	0.0024	0.0024	0.0025	0.0025	0.0025	0.0026	0.0026	0.0027	0.0027	0.0028
30	0.0022	0.0022	0.0023	0.0023	0.0024	0.0024	0.0025	0.0025	0.0025	0.0026	0.0026
34 0	0.0021	0.0021	0.0022	0.0022	0.0022	0.0023	0.0023	0.0024	0.0024	0.0024	0.0025
30	0.0020	0.0020	0.0020	0.0021	0.0021	0.0022	0.0022	0.0022	0.0023	0.0023	0.0023
35 0	0.0018	0.0019	0.0019	0.0020	0.0020	0.0020	0.0021	0.0021	0.0021	0.0022	0.0022
30	0.0017	0.0018	0.0018	0.0018	0.0019	0.0019	0.0019	0.0020	0.0020	0.0020	0.0021
36 0	0.0016	0.0017	0.0017	0.0017	0.0018	0.0018	0.0018	0.0019	0.0019	0.0019	0.0019
30	0.0015	0.0016	0.0016	0.0016	0.0016	0.0017	0.0017	0.0017	0.0018	0.0018	0.0018
37 0	0.0014	0.0014	0.0015	0.0015	0.0015	0.0016	0.0016	0.0016	0.0016	0.0017	0.0017
30	0.0013	0.0013	0.0014	0.0014	0.0014	0.0014	0.0015	0.0015	0.0015	0.0015	0.0016
38 0	0.0012	0.0012	0.0013	0.0013	0.0013	0.0013	0.0014	0.0014	0.0014	0.0014	0.0014
30	0.0011	0.0011	0.0012	0.0012	0.0012	0.0012	0.0012	0.0013	0.0013	0.0013	0.0013
39 0	0.0010	0.0010	0.0011	0.0011	0.0011	0.0011	0.0011	0.0012	0.0012	0.0012	0.0012
30		0.0009	0.0010	0.0010	0.0010	0.0010	0.0010	0.0011	0.0011	0.0011	0.0011
40		0.0008	0.0009	0.0009	0.0009	0.0009	0.0009	0.0009	0.0010	0.0010	0.0010
41		0.0007	0.0007	0.0007	0.0007	0.0007	0.0007	0.0007	0.0007	0.0008	0.0008
42		0.0005	0.0005	0.0005	0.0005	0.0005	0.0005	0.0005	0.0005	0.0005	0.0006
43		0.0003	0.0003	0.0003	0.0003	0.0003	0.0003	0.0003	0.0003	0.0003	0.0004
44		0.0001	0.0001	0.0001	0.0001	0.0001	0.0001	0.0002	0.0002	0.0002	0.0002
45		0.0000	0.0000	0.0000	0.0000	0.0000	0.0000	0.0000	0.0000	0.0000	0.0000
46		9.9998	9.9999	9.9998	9.9998	9.9998	9.9998	9.9998	9.9998	9.9998	9.9998
47		9.9997	9.9997	9.9997	9.9997	9.9996	9.9996	9.9996	9.9996	9.9996	9.9996
48		9.9995	9.9995	9.9995	9.9995	9.9995	9.9995	9.9995	9.9995	9.9994	9.9994
49		9.9994	9.9994	9.9994	9.9993	9.9993	9.9993	9.9993	9.9993	9.9993	9.9993
50		9.9992	9.9992	9.9992	9.9992	9.9992	9.9992	9.9992	9.9991	9.9991	9.9991
51		9.9991	9.9991	9.9991	9.9991	9.9990	9.9990	9.9990	9.9990	9.9990	9.9990
52		9.9990	9.9990	9.9990	9.9989	9.9989	9.9989	9.9989	9.9989	9.9988	9.9988
53		9.9989	9.9988	9.9988	9.9988	9.9988	9.9988	9.9987	9.9987	9.9987	9.9987
54		9.9988	9.9987	9.9987	9.9987	9.9987	9.9986	9.9986	9.9986	9.9986	9.9985
55		9.9986	9.9986	9.9986	9.9986	9.9985	9.9985	9.9985	9.9984	9.9984	9.9984
56		9.9985	9.9985	9.9985	9.9984	9.9984	9.9984	9.9984	9.9983	9.9983	9.9983
57		9.9984	9.9984	9.9984	9.9983	9.9983	9.9983	9.9983	9.9982	9.9982	9.9981
58		9.9983	9.9983	9.9983	9.9982	9.9982	9.9982	9.9982	9.9981	9.9981	9.9980
59		9.9982	9.9982	9.9981	9.9981	9.9981	9.9980	9.9980	9.9980	9.9979	9.9979
60		9.9981	9.9981	9.9980	9.9980	9.9980	9.9979	9.9979	9.9979	9.9978	9.9978
61		9.9980	9.9980	9.9980	9.9979	9.9979	9.9978	9.9978	9.9978	9.9977	9.9977
62		9.9979	9.9979	9.9979	9.9978	9.9978	9.9977	9.9977	9.9977	9.9976	9.9976
63		9.9979	9.9978	9.9978	9.9977	9.9977	9.9976	9.9976	9.9976	9.9975	9.9975
64		9.9978	9.9977	9.9977	9.9976	9.9976	9.9976	9.9975	9.9975	9.9974	9.9974
65		9.9977	9.9977	9.9976	9.9976	9.9975	9.9975	9.9974	9.9974	9.9973	9.9972
66		9.9976	9.9976	9.9975	9.9975	9.9974	9.9974	9.9973	9.9973	9.9973	9.9972
67		9.9976	9.9975	9.9975	9.9974	9.9974	9.9973	9.9973	9.9972	9.9972	9.9971
68		9.9975	9.9974	9.9974	9.9973	9.9973	9.9972	9.9972	9.9971	9.9971	9.9970
69		9.9974	9.9974	9.9973	9.9973	9.9972	9.9972	9.9971	9.9971	9.9970	9.9970
70		9.9974	9.9973	9.9973	9.9972	9.9972	9.9971	9.9971	9.9970	9.9969	9.9969
72		9.9972	9.9972	9.9971	9.9971	9.9970	9.9970	9.9969	9.9969	9.9968	9.9968
74		9.9971	9.9971	9.9970	9.9970	9.9969	9.9969	9.9968	9.9968	9.9967	9.9966
76		9.9971	9.9970	9.9969	9.9969	9.9968	9.9968	9.9967	9.9966	9.9966	9.9965
78		9.9970	9.9969	9.9969	9.9968	9.9967	9.9967	9.9966	9.9966	9.9965	9.9964
80		9.9969	9.9969	9.9968	9.9967	9.9967	9.9966	9.9965	9.9965	9.9964	9.9964
90		9.9968	9.9967	9.9966	9.9966	9.9965	9.9964	9.9964	9.9963	9.9963	9.9962

TABLE VI.

Second Correction of the Lunar Distance.

Apparent Distance	4	7	10	12	14	16	18	20	21	22	23	24	25	26	27	28	29	30	31	32	33	34	35	36	Apparent Distance
Subtr.																									**Add.**
15° 0'	0	2	3	5	6	8	11	13	14	16	17	19	20	22	24	26	27	29	33	33	35	38	40	42	
30	0	2	3	5	6	8	10	13	14	15	17	18	20	21	23	25	26	28	32	32	34	36	39	41	
16 0	0	1	3	4	6	8	10	12	13	15	16	18	19	21	22	24	26	27	31	31	33	35	37	39	
30	0	1	3	4	6	7	9	11	13	14	16	17	18	20	21	23	25	27	30	30	32	34	36	38	
17 0	0	1	3	4	6	7	9	11	13	14	15	16	18	19	21	22	24	26	29	29	31	33	35	37	
30	0	1	3	4	5	7	9	11	12	13	15	16	17	19	20	22	23	25	28	28	30	32	34	36	
18 0	0	1	3	4	5	7	9	11	12	13	14	15	17	18	20	21	23	24	28	28	29	31	33	35	
30	0	1	3	4	5	7	8	10	12	13	14	15	16	18	19	20	22	23	27	27	28	30	32	34	
19 0	0	1	3	4	5	6	8	10	11	12	13	15	16	17	18	20	21	23	26	26	28	29	31	33	
30	0	1	2	4	5	6	8	10	11	12	13	14	15	17	18	19	21	22	25	25	27	28	30	32	
20 0	0	1	2	3	5	6	8	10	11	12	13	14	15	16	17	19	20	22	25	25	26	28	29	31	
21	0	1	2	3	5	6	7	9	10	11	12	13	14	15	17	18	19	20	23	23	25	26	28	29	
22	0	1	2	3	4	6	7	9	10	10	11	12	14	15	16	17	18	19	22	22	24	25	26	28	
23	0	1	2	3	4	5	7	8	9	10	11	12	13	14	15	16	17	19	21	21	22	24	25	27	
24	0	1	2	3	4	5	6	8	9	9	10	11	13	14	15	15	16	18	20	20	21	23	24	25	
25	0	1	2	3	4	5	6	7	8	9	10	11	12	13	14	15	16	17	19	19	20	22	23	24	
26	0	1	2	3	4	5	6	7	8	9	9	10	11	12	13	14	15	16	18	18	19	21	22	23	
27	0	1	2	2	3	4	6	7	8	8	9	10	11	12	12	13	14	15	18	18	19	20	21	22	
28	0	1	2	2	3	4	5	7	7	8	9	9	10	11	12	13	14	15	17	17	18	19	20	21	
29	0	1	2	2	3	4	5	6	7	8	8	9	10	11	11	12	13	14	16	16	17	18	19	20	
30	0	1	2	2	3	4	5	6	7	7	8	9	9	10	11	12	13	14	15	15	16	17	19	20	
31	0	1	1	2	3	4	5	6	6	7	8	8	9	10	11	11	12	13	15	15	16	17	18	19	
32	0	1	1	2	3	4	5	6	6	7	7	8	9	9	10	11	12	13	14	14	15	16	17	18	
33	0	1	1	2	3	3	4	5	6	7	7	8	8	9	10	11	11	12	14	14	15	16	16	17	
34	0	1	1	2	3	3	4	5	6	6	7	7	8	9	9	10	11	12	13	13	14	15	16	17	
35	0	1	1	2	2	3	4	5	5	6	7	7	8	8	9	10	10	11	13	13	14	14	15	16	
36	0	1	1	2	2	3	4	5	5	6	6	7	7	8	8	9	10	11	12	12	13	14	15	16	
37	0	1	1	2	2	3	4	5	5	6	6	7	7	8	8	9	10	10	12	12	13	13	14	15	
38	0	1	1	2	2	3	4	4	5	5	6	6	7	8	8	9	9	10	11	11	12	13	14	14	
39	0	1	1	2	2	3	3	4	5	5	6	6	7	7	8	8	9	10	11	11	12	12	13	14	
40	0	1	1	2	2	3	3	4	5	5	6	6	7	7	8	8	9	9	11	11	11	12	13	13	140°
42	0	0	1	1	2	2	3	4	4	5	5	6	6	7	7	8	8	9	10	10	11	11	12	13	138
44	0	0	1	1	2	2	3	4	4	4	5	5	6	6	7	7	8	8	9	9	10	10	11	12	136
46	0	0	1	1	2	2	3	3	4	4	4	5	5	6	6	7	7	8	9	9	9	10	10	11	134
48	0	0	1	1	2	2	3	3	3	4	4	5	5	5	6	6	7	7	8	8	9	9	10	11	132
50	0	0	1	1	1	2	2	3	3	4	4	4	5	5	5	6	6	7	8	8	8	8	9	9	130
52	0	0	1	1	1	2	2	3	3	3	4	4	4	5	5	5	6	6	7	7	7	8	8	9	128
54	0	0	1	1	1	2	2	3	3	3	3	4	4	4	5	5	5	6	6	6	7	7	8	8	126
56	0	0	1	1	1	2	2	2	2	3	3	3	4	4	4	5	5	6	6	6	6	7	7	8	124
58	0	0	1	1	1	1	2	2	2	3	3	3	3	4	4	4	5	5	6	6	6	6	7	7	122
60	0	0	0	1	1	1	2	2	2	2	3	3	3	3	4	4	4	5	5	5	5	6	6	7	120
62	0	0	0	1	1	1	2	2	2	2	2	3	3	3	3	4	4	4	5	5	5	5	6	6	118
64	0	0	0	1	1	1	1	2	2	2	2	2	3	3	3	3	4	4	4	4	4	4	5	6	116
66	0	0	0	1	1	1	1	2	2	2	2	2	2	3	3	3	3	4	4	4	4	4	5	5	114
68	0	0	0	1	1	1	1	1	2	2	2	2	2	2	3	3	3	3	4	4	4	4	4	5	112
70	0	0	0	0	1	1	1	1	1	2	2	2	2	2	2	2	3	3	3	3	3	4	4	4	110
74	0	0	0	0	0	1	1	1	1	1	1	1	2	2	2	2	2	2	3	3	3	3	3	3	106
78	0	0	0	0	0	0	1	1	1	1	1	1	1	1	1	1	2	2	2	2	2	2	2	2	102
82	0	0	0	0	0	0	0	0	1	1	1	1	1	1	1	1	1	1	1	1	1	1	2	2	98
86	0	0	0	0	0	0	0	0	0	0	0	0	0	0	0	0	1	1	1	1	1	1	1	1	94
90°	0	0	0	0	0	0	0	0	0	0	0	0	0	0	0	0	0	0	0	0	0	0	0	0	90°

FIRST CORRECTION OF DISTANCE.

TABLE VI.

Second Correction of the Lunar Distance.

Apparent Distance.	\multicolumn{24}{c}{FIRST CORRECTION OF DISTANCE.}	Apparent Distance.
	37	38	39	40	41	42	43	44	45	46	47	48	49	50	51	52	53	54	55	56	57	58	59	60	
Subtr 15° 0′	45	47	50	52	55	57	60	63	66	69	72	75	78	81	85	88	91	95	99	102	106	110	113	117	A.M.
30	43	45	48	50	53	56	58	61	64	67	70	72	76	79	82	85	88	92	95	99	102	106	110	113	
16 0	42	44	46	49	51	54	56	59	62	64	67	70	73	76	79	82	85	89	92	95	99	102	106	110	
30	40	43	45	47	50	52	54	57	60	62	65	68	71	74	77	80	83	86	89	92	96	99	103	106	
17 0	39	41	43	46	48	50	53	55	58	60	63	66	69	71	74	77	80	83	86	90	93	96	99	103	
30	38	40	42	44	47	49	51	54	56	59	61	64	66	69	72	75	78	81	84	87	90	93	96	100	
18 0	37	39	41	43	45	47	50	52	54	57	59	62	64	67	70	73	75	78	81	84	87	90	94	97	
30	36	38	40	42	44	46	48	50	53	55	58	60	63	65	68	71	73	76	79	82	85	88	91	94	
19 0	35	37	39	41	43	45	47	49	51	54	56	58	61	63	66	69	71	74	77	79	82	85	88	91	
30	34	36	37	39	41	43	46	48	50	52	54	57	59	62	64	67	69	72	75	77	80	83	86	89	
20	33	35	36	38	40	42	44	46	49	51	53	55	58	60	62	65	67	70	73	75	78	81	83	86	
21	31	33	35	36	38	40	42	44	46	48	50	52	55	57	59	61	64	66	69	71	74	76	79	82	
22	30	31	33	35	36	38	40	42	44	46	48	50	52	54	56	58	61	63	65	68	70	73	75	78	
23	28	30	31	33	35	36	38	40	42	44	45	47	49	51	53	56	58	60	62	64	67	69	72	74	
24	27	28	30	31	33	35	36	38	40	41	43	45	47	49	51	53	55	57	59	61	64	66	68	71	
25	26	27	28	30	31	33	35	36	38	40	41	43	45	47	49	51	53	55	57	59	61	63	65	67	
26	25	26	27	29	30	32	33	35	36	38	40	41	43	45	47	48	50	52	54	56	58	60	62	64	
27	23	25	26	27	29	30	32	33	35	36	38	39	41	43	45	46	48	50	52	54	56	58	60	62	
28	22	24	25	26	28	29	30	32	33	35	36	38	39	41	43	44	46	48	50	51	53	55	57	59	
29	22	23	24	25	26	28	29	30	32	33	35	36	38	39	41	43	44	46	48	49	51	53	55	57	
30	21	22	23	24	25	27	28	29	31	32	33	35	36	38	39	41	42	44	46	47	49	51	53	54	
31	20	21	22	23	24	26	27	28	29	31	32	33	35	36	38	39	41	42	44	46	47	49	51	52	
32	19	20	21	22	23	25	26	27	28	30	31	32	34	35	36	38	39	41	42	44	45	47	49	50	
33	18	19	20	22	23	24	25	26	27	28	30	31	32	34	35	36	38	39	41	42	44	45	47	48	
34	18	19	20	21	22	23	24	25	26	27	29	30	31	32	34	35	36	38	39	41	42	44	45	47	
35	17	18	19	20	21	22	23	24	25	26	28	29	30	31	32	34	35	36	38	39	40	42	43	45	
36	16	17	18	19	20	21	22	23	24	25	27	28	29	30	31	32	34	35	36	38	39	40	42	43	
37	16	17	18	19	19	20	21	22	23	25	26	27	28	29	30	31	33	34	35	36	38	39	40	42	
38	15	16	17	18	19	20	20	21	22	23	24	25	26	27	29	30	31	33	34	35	36	38	39	40	
39	15	16	16	17	18	19	20	21	22	23	24	25	26	27	28	29	30	31	33	34	35	36	38	39	
40	14	15	16	17	17	18	19	20	21	22	23	24	25	26	27	28	29	30	31	33	34	35	36	37	140°
42	13	14	15	16	16	17	18	19	20	21	21	22	23	24	25	26	27	28	29	30	31	33	34	35	138
44	12	13	14	14	15	16	17	17	18	19	20	21	22	23	24	24	25	26	27	28	29	30	31	33	136
46	12	12	13	13	14	15	16	16	17	18	19	19	20	21	22	23	24	25	26	26	27	28	29	30	134
48	11	11	12	13	13	14	15	15	16	17	17	18	19	20	20	21	22	23	24	25	26	26	27	28	132
50	10	11	11	12	12	13	14	14	15	16	16	17	18	18	19	20	21	21	22	23	24	25	25	26	130
52	9	10	10	11	11	12	13	13	14	14	15	15	16	17	18	18	19	20	21	21	22	23	24	25	128
54	9	9	10	10	11	11	12	12	13	13	14	15	15	16	16	17	18	18	19	20	21	21	22	23	126
56	8	9	9	9	10	10	11	11	12	12	13	14	14	15	15	16	17	17	18	18	19	20	20	21	124
58	7	8	8	9	9	10	10	11	11	12	12	13	13	14	14	15	15	16	16	17	18	18	19	20	122
60	7	7	8	8	8	9	9	10	10	11	11	12	12	13	13	14	15	15	16	17	18	18	18	18	120
62	6	7	7	7	8	8	9	9	9	10	10	11	11	12	12	13	13	14	14	15	15	16	16	17	118
64	6	6	6	7	7	8	8	8	9	9	9	10	10	11	11	12	12	12	13	13	14	14	15	15	116
66	5	6	6	6	7	7	7	8	8	8	9	9	9	10	10	11	11	11	12	12	13	13	14	14	114
68	5	5	5	6	6	6	7	7	7	7	8	8	8	9	9	10	10	10	11	11	11	12	12	13	112
70	4	5	5	5	5	6	6	6	6	7	7	7	8	8	8	9	9	9	10	10	10	11	11	11	110
74	3	4	4	4	4	4	5	5	5	5	6	6	6	6	7	7	7	7	8	8	8	8	9	9	106
78	3	3	3	3	3	3	4	4	4	4	4	5	5	5	6	6	5	5	6	6	6	6	6	7	102
82	2	2	2	2	2	2	2	3	3	3	3	3	3	3	3	3	3	4	4	4	4	4	4	4	98
86	1	1	1	1	1	1	1	1	1	1	1	1	1	2	2	2	2	2	2	2	2	2	2	2	94
90°	0	0	0	0	0	0	0	0	0	0	0	0	0	0	0	0	0	0	0	0	0	0	0	0	90°
Apparent Distance.	37	38	39	40	41	42	43	44	45	46	47	48	49	50	51	52	53	54	55	56	57	58	59	60	Apparent Distance.

FIRST CORRECTION OF DISTANCE.

TABLE VII.

For finding the Correction of the Lunar Distance for the Contraction of the Moon's Semidiameter.

TABLE VII. A. GIVING THE ARGUMENT FOR TABLE VII. B.

Reduced P. and R of ☽	APPARENT ALTITUDE OF ☽																								
	5	5½	6	6½	7	7½	8	8½	9	9½	10	11	12	13	14	15	16	17	18	20	25	30	40	50
41′	65	56																						
42	63	54	47	41																				
43	62	53	46	40	35																			
44	60	51	45	39	34	30	27																	
45	58	50	43	38	33	30	26	24	21	20														
46	57	49	42	37	33	29	26	23	21	19	17	15												
47	56	48	41	36	32	28	25	23	20	19	17	14	12	10										
48	54	46	40	35	31	28	25	22	20	18	17	14	12	10	9	8	7	6						
49	53	45	39	35	30	27	24	22	19	18	16	14	12	10	9	8	7	6	6	5	3			
50	52	44	38	34	30	26	24	21	19	17	16	13	11	10	9	8	7	6	5	5	3	3	2	
51	50	43	38	33	29	26	23	21	19	17	15	13	11	10	8	7	7	6	5	5	3	2	2	2
52	49	42	37	32	28	25	23	20	18	17	15	13	11	9	8	7	6	5	4	3	2	2	2	2
53	48	41	36	32	28	25	22	20	18	16	15	12	11	9	8	7	6	6	5	4	3	2	2	2
54	47	41	35	31	27	24	22	19	18	16	15	12	10	9	8	7	6	6	5	4	3	2	2	2
55			35	30	27	24	21	19	17	16	14	12	10	9	8	7	6	6	5	4	3	2	2	2
56					26	23	21	19	17	15	14	12	10	9	8	7	6	5	5	4	3	2	2	2
57						18	17	15	14	12	10	9	7	7	6	5	5	4	3	2	2	2		
58									13	11	10	8	7	6	5	5	4	3	2	2	2			
59												8	7	6	6	5	5	4	3	2	2	2		
60																	4	3	2	2	2			

TABLE VII. B. CONTRACTION OF ☽'S SEMIDIAMETER.

Whole Correction of ☽	ARGUMENT = NUMBER FROM TABLE VII. A.																															
	2	4	6	8	10	12	14	16	18	20	22	24	26	28	30	32	34	36	38	40	44	48	52	56	60	64
0′	0″	0″	0″	0″	0″	0″	0″	0″	0″	0″	0″	0″	0″	0″	0″	0″	0″	0″	0″	0″	0″	0″	0″	0″	0″	0″
5	0	0	0	0	0	0	0	0	0	0	0	0	0	0	0	0	0	0	0	0	0	0	0	0	0	0
10	0	0	0	0	0	0	0	0	0	0	0	0	1	1	1	1	1	1	1	1	1	1	1	1	1	1
15	0	0	0	0	0	1	1	1	1	1	1	1	1	1	1	1	2	2	2	2	2	2	2	3	3	3
20	0	0	0	1	1	1	1	1	1	2	2	2	2	2	2	3	3	3	3	3	4	4	4	4	5	5
22	0	0	1	1	1	1	1	2	2	2	2	2	3	3	3	3	3	3	4	4	4	5	5	5	6	6
24	0	0	1	1	1	1	2	2	2	2	3	3	3	3	3	4	4	4	4	5	5	6	6	6	7	7
26	0	1	1	1	1	2	2	2	2	3	3	3	3	4	4	5	5	5	5	6	6	6	7	8	8	9
28	0	1	1	1	2	2	2	3	3	3	3	4	4	5	5	5	6	6	6	7	8	8	9	9	10	
30	0	1	1	1	2	2	3	3	3	4	4	5	5	5	6	6	6	7	7	8	9	9	10	11	12	
32	0	1	1	2	2	2	3	3	4	4	5	5	6	6	7	7	7	8	8	9	10	11	11	12	13	
34	0	1	1	2	2	3	3	4	4	5	5	6	6	7	8	8	9	9	10	11	12	13	14	15		
36	1	1	2	2	3	3	4	4	5	5	6	6	7	7	8	8	9	9	10	10	11	12	13	15	16	17
38	1	1	2	2	3	3	4	5	5	6	6	7	8	8	9	9	10	10	11	12	13	14	15	16	17	18
40	1	1	2	3	3	4	4	5	6	6	7	8	8	9	9	10	11	12	12	13	14	15	17	18	19	20
42	1	1	2	3	4	4	5	6	6	7	8	8	9	10	11	11	12	13	13	14	16	17	18	20	21	23
44	1	2	2	3	4	5	5	6	7	8	9	9	10	11	12	12	13	14	15	15	17	19	20	22	23	
45	1	2	2	3	4	5	6	6	7	8	9	10	11	11	12	13	14	15	15	16	18	19	21	23	24	
46	1	2	3	3	4	5	6	7	8	9	10	11	12	13	14	14	15	16	17	17	19	20	22	24		
47	1	2	3	4	4	5	6	7	8	9	10	11	11	12	13	14	15	16	17	18	19	21	23	25		
48	1	2	3	4	5	6	6	7	8	9	10	11	12	13	14	15	16	17	18	18	20	22	24	26		
49	1	2	3	4	5	6	7	8	9	10	11	12	12	13	14	15	16	17	18	19	21	23	25			
50	1	2	3	4	5	6	7	8	9	10	11	12	13	14	15	16	17	18	19	20	22	24	26			
51	1	2	3	4	5	6	7	8	9	10	11	12	14	15	16	17	18	19	20	21	23	25	27			
52	1	2	3	4	5	6	8	9	10	11	12	13	14	15	16	17	18	19	21	22	24	26				
53	1	2	3	4	6	7	8	9	10	11	12	13	15	16	17	18	19	20	21	22	25	27				
54		2	3	5	6	7	8	9	10	12	13	14	15	16	17	19	20	21	22	23	26					
55		2	4	5	6	7	8	10	11	12	13	15	16	17	18	19	21	22								
56		3	4	5	6	8	9	10	11	13	14	15	16													
57			4	5	7																					

When the *nearest* limb is used *subtract* this correction; when the *farthest*, add.

33

TABLE VIII.

For finding the Correction of the Lunar Distance for the Contraction of the Sun's Semidiameter.

TABLE VIII. A. GIVING THE ARGUMENT FOR TABLE VIII. B.

Reduced P. and R. of ☉.	APPARENT ALTITUDE OF ☉.																						
	5	5½	6	6½	7	7¼	8	8½	9	9½	10	11	12	13	14	15	16	17	18 20	25 30	40 50		
1′ 0″																					22 18		
30																				30 34	24 29		
2 0																			35 37	42 46	46		
30																	40 42	44 47	53 59			
3 0															44 46	49 51	53 57					
30													45 48	51 54	57 60	62 67						
4 0										45 49	52 55	59 62	65 68									
30									47 49	51 55	59 63	66 70										
5 0							47 50	52 54	57 61	66 70	74											
30					47 50	52 55	57 60	62 67	72													
6 0				49	52 55	57 60	63 66	68 74														
30			50 53	56 59	62 65	68 71	74															
7 0		51	54 58	61 64	67 70	74																
30		55 58	62	65 69	72 75																	
8 0	55 59	62 66	70 73	77																		
30	59 63	66 70	74 78																			
9 0	62 66	70 74	79																			
30	66 70	74 79																				
10 0	69 74	78																				
30	73 77																					
11 0	76 81																					
30	80																					

TABLE VIII. B. CONTRACTION OF ☉'S SEMIDIAMETER.

Whole Correction of ☉.	ARGUMENT = NUMBER FROM TABLE VIII. A.																			
	20	24	28	32	36 40	44 46	48 50	52 54	56 58	60 62	64 66	68 70	72 74	76 78						
0′ 0″	0	0	0	0	0 0	0 0	0 0	0 0	0 0	0 0	0 0	0 0	0 0	0 0						
1 0	1	1	1	1	1 0	0 0	0 0	0 0	0 0	0 0	0 0	0 0	0 0	0 0						
2 0			2	2	2 2	2 2	2 1	1 1	1 1	1 1	1 1	1 1	1 1	1 1						
30				3	3 3	3 2	2 2	2 2	2 2	2 2	2 2	2 2	2 2	2 2						
3 0					4 4	4 4	3 3	3 3	3 3	3 3	3 2	2 2	2 2	2 2						
30						5 5	5 4	4 4	4 4	4 4	3 3	3 3	3 3	3 3						
4 0						7 6	6 6	6 5	5 5	5 5	5 4	4 4	4 4	4 4						
20							7	7 7	7 6	6 6	6 5	5 5	5 5	4 4						
40							9	8 8	8 7	7 7	6 6	6 6	5 5	5 5						
5 0							10	9 9	9 8	8 8	8 7	7 7	6 6	6 6						
20								11 10	10 9	9 9	9 8	8 8	8 7	7 7	7 7					
40								12 12	11 11	10 10	10 9	9 9	9 8	8 8	8 7					
6 0								13	12 12	12 11	11 10	10 10	10 9	9 9	9 8					
20								14 14	13 13	12 12	12 12	11 11	11 10	10 10	10 9					
40								16 15	15 14	14 13	13 13	12 12	11 11	11 10						
7 0								18	17 16	16 15	15 14	14 13	13 13	12 12	12 11					
20								19 18	17 17	16 16	15 15	14 14	13 13	13 12						
40								20	19 18	18 17	17 16	16 15	15 14	14 14						
8 0								21	21 20	19 19	18 17	17 16	16 16	15 15						
20									22	21 20	20 19	18 18	17 17	16 16						
40									23	23 22	21 20	20 19	19 18	18 17						
9 0									24	23 22	21 21	20 20	19 19							
20										25 24	23 22	22 21	21 20							
40										25	25 24	23 23	22 22							
10 0											26 26	25 24	24 23							
20											28	27 26	25 25							
40												28 28	27 26							
11 0													29 28							
20													30							

Subtract this correction from the distance.

39

TABLE IX.

LOGARITHMS OF SMALL ARCS IN SPACE OR TIME.

Arc. (ʰ ᵐ ˢ)	0″	1″	2″	3″	4″	5″	6″	7″	8″	9″
0 0 0		0.0000	0.3010	0.4771	0.6021	0.6990	0.7782	0.8451	0.9031	0.9542
0 0 10	1.0000	1.0414	1.0792	1.1139	1.1461	1.1761	1.2041	1.2304	1.2553	1.2788
0 0 20	1.3010	1.3222	1.3424	1.3617	1.3802	1.3979	1.4150	1.4314	1.4472	1.4624
0 0 30	1.4771	1.4914	1.5051	1.5185	1.5315	1.5441	1.5563	1.5682	1.5798	1.5911
0 0 40	1.6021	1.6128	1.6232	1.6335	1.6435	1.6532	1.6628	1.6721	1.6812	1.6902
0 0 50	1.6990	1.7076	1.7160	1.7243	1.7324	1.7404	1.7482	1.7559	1.7634	1.7709
0 1 0	1.7782	1.7853	1.7924	1.7993	1.8062	1.8129	1.8195	1.8261	1.8325	1.8388
1 10	1.8451	1.8513	1.8573	1.8633	1.8692	1.8751	1.8808	1.8865	1.8921	1.8976
1 20	1.9031	1.9085	1.9138	1.9191	1.9243	1.9294	1.9345	1.9395	1.9445	1.9494
1 30	1.9542	1.9589	1.9638	1.9685	1.9731	1.9777	1.9823	1.9868	1.9912	1.9956
1 40	2.0000	2.0043	2.0086	2.0128	2.0170	2.0212	2.0253	2.0294	2.0334	2.0374
1 50	2.0414	2.0453	2.0492	2.0531	2.0569	2.0607	2.0645	2.0682	2.0719	2.0755
0 2 0	2.0792	2.0828	2.0864	2.0899	2.0934	2.0969	2.1004	2.1038	2.1072	2.1106
2 10	2.1139	2.1173	2.1206	2.1239	2.1271	2.1303	2.1335	2.1367	2.1399	2.1430
2 20	2.1461	2.1492	2.1523	2.1553	2.1584	2.1614	2.1644	2.1673	2.1703	2.1732
2 30	2.1761	2.1790	2.1818	2.1847	2.1875	2.1903	2.1931	2.1959	2.1987	2.2014
2 40	2.2041	2.2068	2.2095	2.2122	2.2148	2.2175	2.2201	2.2227	2.2253	2.2279
2 50	2.2304	2.2330	2.2355	2.2380	2.2405	2.2430	2.2455	2.2480	2.2504	2.2529
0 3 0	2.2553	2.2577	2.2601	2.2625	2.2648	2.2672	2.2695	2.2718	2.2742	2.2765
3 10	2.2788	2.2810	2.2833	2.2856	2.2878	2.2900	2.2923	2.2945	2.2967	2.2989
3 20	2.3010	2.3032	2.3054	2.3075	2.3096	2.3118	2.3139	2.3160	2.3181	2.3201
3 30	2.3222	2.3243	2.3263	2.3284	2.3304	2.3324	2.3345	2.3365	2.3385	2.3404
3 40	2.3424	2.3444	2.3464	2.3483	2.3502	2.3522	2.3541	2.3560	2.3579	2.3598
3 50	2.3617	2.3636	2.3655	2.3674	2.3692	2.3711	2.3729	2.3747	2.3766	2.3784
0 4 0	2.3802	2.3820	2.3838	2.3856	2.3874	2.3892	2.3909	2.3927	2.3945	2.3962
4 10	2.3979	2.3997	2.4014	2.4031	2.4048	2.4065	2.4082	2.4099	2.4116	2.4133
4 20	2.4150	2.4166	2.4183	2.4200	2.4216	2.4232	2.4249	2.4265	2.4281	2.4298
4 30	2.4314	2.4330	2.4346	2.4362	2.4378	2.4393	2.4409	2.4425	2.4440	2.4456
4 40	2.4472	2.4487	2.4502	2.4518	2.4533	2.4548	2.4564	2.4579	2.4594	2.4609
4 50	2.4624	2.4639	2.4654	2.4669	2.4683	2.4698	2.4713	2.4728	2.4742	2.4757
0 5 0	2.4771	2.4786	2.4800	2.4814	2.4829	2.4843	2.4857	2.4871	2.4886	2.4900
5 10	2.4914	2.4928	2.4942	2.4955	2.4969	2.4983	2.4997	2.5011	2.5024	2.5038
5 20	2.5051	2.5065	2.5079	2.5092	2.5105	2.5119	2.5132	2.5145	2.5159	2.5172
5 30	2.5185	2.5198	2.5211	2.5224	2.5237	2.5250	2.5263	2.5276	2.5289	2.5302
5 40	2.5315	2.5328	2.5340	2.5353	2.5366	2.5378	2.5391	2.5403	2.5416	2.5428
5 50	2.5441	2.5453	2.5465	2.5478	2.5490	2.5502	2.5514	2.5527	2.5539	2.5551
0 6 0	2.5563	2.5575	2.5587	2.5599	2.5611	2.5623	2.5635	2.5647	2.5658	2.5670
6 10	2.5682	2.5694	2.5705	2.5717	2.5729	2.5740	2.5752	2.5763	2.5775	2.5786
6 20	2.5798	2.5809	2.5821	2.5832	2.5843	2.5855	2.5866	2.5877	2.5888	2.5899
6 30	2.5911	2.5922	2.5933	2.5944	2.5955	2.5966	2.5977	2.5988	2.5999	2.6010
6 40	2.6021	2.6031	2.6042	2.6053	2.6064	2.6075	2.6085	2.6096	2.6107	2.6117
6 50	2.6128	2.6138	2.6149	2.6160	2.6170	2.6180	2.6191	2.6201	2.6212	2.6222
0 7 0	2.6232	2.6243	2.6253	2.6263	2.6274	2.6284	2.6294	2.6304	2.6314	2.6325
7 10	2.6335	2.6345	2.6355	2.6365	2.6375	2.6385	2.6395	2.6405	2.6415	2.6425
7 20	2.6435	2.6444	2.6454	2.6464	2.6474	2.6484	2.6493	2.6503	2.6513	2.6522
7 30	2.6532	2.6542	2.6551	2.6561	2.6571	2.6580	2.6590	2.6599	2.6609	2.6618
7 40	2.6628	2.6637	2.6646	2.6656	2.6665	2.6675	2.6684	2.6693	2.6702	2.6712
7 50	2.6721	2.6730	2.6739	2.6749	2.6758	2.6767	2.6776	2.6785	2.6794	2.6803
0 8 0	2.6812	2.6821	2.6830	2.6839	2.6848	2.6857	2.6866	2.6875	2.6884	2.6893
8 10	2.6902	2.6911	2.6920	2.6928	2.6937	2.6946	2.6955	2.6964	2.6972	2.6981
8 20	2.6990	2.6998	2.7007	2.7016	2.7024	2.7033	2.7042	2.7050	2.7059	2.7067
8 30	2.7076	2.7084	2.7093	2.7101	2.7110	2.7118	2.7126	2.7135	2.7143	2.7152
8 40	2.7160	2.7168	2.7177	2.7185	2.7193	2.7202	2.7210	2.7218	2.7226	2.7235
8 50	2.7243	2.7251	2.7259	2.7267	2.7275	2.7284	2.7292	2.7300	2.7308	2.7316
0 9 0	2.7324	2.7332	2.7340	2.7348	2.7356	2.7364	2.7372	2.7380	2.7388	2.7396
9 10	2.7404	2.7412	2.7419	2.7427	2.7435	2.7443	2.7451	2.7459	2.7466	2.7474
9 20	2.7482	2.7490	2.7497	2.7505	2.7513	2.7520	2.7528	2.7536	2.7543	2.7551
9 30	2.7559	2.7566	2.7574	2.7582	2.7589	2.7597	2.7604	2.7612	2.7619	2.7627
9 40	2.7634	2.7642	2.7649	2.7657	2.7664	2.7672	2.7679	2.7686	2.7694	2.7701
9 50	2.7709	2.7716	2.7723	2.7731	2.7738	2.7745	2.7752	2.7760	2.7767	2.7774

TABLE IX.

LOGARITHMS OF SMALL ARCS IN SPACE OR TIME.

Arc.	0″	1″	2″	3″	4″	5″	6″	7″	8″	9″
0ʰ·10ᵐ·0ˢ	2.7782	2.7789	2.7796	2.7803	2.7810	2.7818	2.7825	2.7832	2.7839	2.7846
10 10	2.7853	2.7860	2.7868	2.7875	2.7882	2.7889	2.7896	2.7903	2.7910	2.7917
10 20	2.7924	2.7931	2.7938	2.7945	2.7952	2.7959	2.7966	2.7973	2.7980	2.7987
10 30	2.7993	2.8000	2.8007	2.8014	2.8021	2.8028	2.8035	2.8041	2.8048	2.8055
10 40	2.8062	2.8069	2.8075	2.8082	2.8089	2.8096	2.8102	2.8109	2.8116	2.8122
10 50	2.8129	2.8136	2.8142	2.8149	2.8156	2.8162	2.8169	2.8176	2.8182	2.8189
0 11 0	2.8195	2.8202	2.8209	2.8215	2.8222	2.8228	2.8235	2.8241	2.8248	2.8254
11 10	2.8261	2.8267	2.8274	2.8280	2.8287	2.8293	2.8299	2.8306	2.8312	2.8319
11 20	2.8325	2.8331	2.8338	2.8344	2.8351	2.8357	2.8363	2.8370	2.8376	2.8382
11 30	2.8388	2.8395	2.8401	2.8407	2.8414	2.8420	2.8426	2.8432	2.8439	2.8445
11 40	2.8451	2.8457	2.8463	2.8470	2.8476	2.8482	2.8488	2.8494	2.8500	2.8506
11 50	2.8513	2.8519	2.8525	2.8531	2.8537	2.8543	2.8549	2.8555	2.8561	2.8567
0 12 0	2.8573	2.8579	2.8585	2.8591	2.8597	2.8603	2.8609	2.8615	2.8621	2.8627
12 10	2.8633	2.8639	2.8645	2.8651	2.8657	2.8663	2.8669	2.8675	2.8681	2.8686
12 20	2.8692	2.8698	2.8704	2.8710	2.8716	2.8722	2.8727	2.8733	2.8739	2.8745
12 30	2.8751	2.8756	2.8762	2.8768	2.8774	2.8779	2.8785	2.8791	2.8797	2.8802
12 40	2.8808	2.8814	2.8820	2.8825	2.8831	2.8837	2.8842	2.8848	2.8854	2.8859
12 50	2.8865	2.8871	2.8876	2.8882	2.8887	2.8893	2.8899	2.8904	2.8910	2.8915
0 13 0	2.8921	2.8927	2.8932	2.8938	2.8943	2.8949	2.8954	2.8960	2.8965	2.8971
13 10	2.8976	2.8982	2.8987	2.8993	2.8998	2.9004	2.9009	2.9015	2.9020	2.9025
13 20	2.9031	2.9036	2.9042	2.9047	2.9053	2.9058	2.9063	2.9069	2.9074	2.9079
13 30	2.9085	2.9090	2.9096	2.9101	2.9106	2.9112	2.9117	2.9122	2.9128	2.9133
13 40	2.9138	2.9143	2.9149	2.9154	2.9159	2.9165	2.9170	2.9175	2.9180	2.9186
13 50	2.9191	2.9196	2.9201	2.9206	2.9212	2.9217	2.9222	2.9227	2.9232	2.9238
0 14 0	2.9243	2.9248	2.9253	2.9258	2.9263	2.9269	2.9274	2.9279	2.9284	2.9289
14 10	2.9294	2.9299	2.9304	2.9309	2.9315	2.9320	2.9325	2.9330	2.9335	2.9340
14 20	2.9345	2.9350	2.9355	2.9360	2.9365	2.9370	2.9375	2.9380	2.9385	2.9390
14 30	2.9395	2.9400	2.9405	2.9410	2.9415	2.9420	2.9425	2.9430	2.9435	2.9440
14 40	2.9445	2.9450	2.9455	2.9460	2.9465	2.9469	2.9474	2.9479	2.9484	2.9489
14 50	2.9494	2.9499	2.9504	2.9509	2.9513	2.9518	2.9523	2.9528	2.9533	2.9538
0 15 0	2.9542	2.9547	2.9552	2.9557	2.9562	2.9566	2.9571	2.9576	2.9581	2.9586
15 10	2.9590	2.9595	2.9600	2.9605	2.9609	2.9614	2.9619	2.9624	2.9628	2.9633
15 20	2.9638	2.9643	2.9647	2.9652	2.9657	2.9661	2.9666	2.9671	2.9675	2.9680
15 30	2.9685	2.9689	2.9694	2.9699	2.9703	2.9708	2.9713	2.9717	2.9722	2.9727
15 40	2.9731	2.9736	2.9741	2.9745	2.9750	2.9754	2.9759	2.9763	2.9768	2.9773
15 50	2.9777	2.9782	2.9786	2.9791	2.9795	2.9800	2.9805	2.9809	2.9814	2.9818
0 16 0	2.9823	2.9827	2.9832	2.9836	2.9841	2.9845	2.9850	2.9854	2.9859	2.9863
16 10	2.9868	2.9872	2.9877	2.9881	2.9886	2.9890	2.9894	2.9899	2.9903	2.9908
16 20	2.9912	2.9917	2.9921	2.9926	2.9930	2.9934	2.9939	2.9943	2.9948	2.9952
16 30	2.9956	2.9961	2.9965	2.9969	2.9974	2.9978	2.9983	2.9987	2.9991	2.9996
16 40	3.0000	3.0004	3.0009	3.0013	3.0017	3.0022	3.0026	3.0030	3.0035	3.0039
16 50	3.0043	3.0048	3.0052	3.0056	3.0060	3.0065	3.0069	3.0073	3.0077	3.0082
0 17 0	3.0086	3.0090	3.0095	3.0099	3.0103	3.0107	3.0111	3.0116	3.0120	3.0124
17 10	3.0128	3.0133	3.0137	3.0141	3.0145	3.0149	3.0154	3.0158	3.0162	3.0166
17 20	3.0170	3.0175	3.0179	3.0183	3.0187	3.0191	3.0195	3.0199	3.0204	3.0208
17 30	3.0212	3.0216	3.0220	3.0224	3.0228	3.0233	3.0237	3.0241	3.0245	3.0249
17 40	3.0253	3.0257	3.0261	3.0265	3.0269	3.0273	3.0278	3.0282	3.0286	3.0290
17 50	3.0294	3.0298	3.0302	3.0306	3.0310	3.0314	3.0318	3.0322	3.0326	3.0330
0 18 0	3.0334	3.0338	3.0342	3.0346	3.0350	3.0354	3.0358	3.0362	3.0366	3.0370
18 10	3.0374	3.0378	3.0382	3.0386	3.0390	3.0394	3.0398	3.0402	3.0406	3.0410
18 20	3.0414	3.0418	3.0422	3.0426	3.0430	3.0434	3.0438	3.0441	3.0445	3.0449
18 30	3.0453	3.0457	3.0461	3.0465	3.0469	3.0473	3.0477	3.0481	3.0484	3.0488
18 40	3.0492	3.0496	3.0500	3.0504	3.0508	3.0512	3.0515	3.0519	3.0523	3.0527
18 50	3.0531	3.0535	3.0538	3.0542	3.0546	3.0550	3.0554	3.0558	3.0561	3.0565
0 19 0	3.0569	3.0573	3.0577	3.0580	3.0584	3.0588	3.0592	3.0596	3.0599	3.0603
19 10	3.0607	3.0611	3.0615	3.0618	3.0622	3.0626	3.0630	3.0633	3.0637	3.0641
19 20	3.0645	3.0648	3.0652	3.0656	3.0660	3.0663	3.0667	3.0671	3.0674	3.0678
19 30	3.0682	3.0686	3.0689	3.0693	3.0697	3.0700	3.0704	3.0708	3.0711	3.0715
19 40	3.0719	3.0722	3.0726	3.0730	3.0734	3.0737	3.0741	3.0745	3.0748	3.0752
19 50	3.0755	3.0759	3.0763	3.0766	3.0770	3.0774	3.0777	3.0781	3.0785	3.0788

TABLE IX.

LOGARITHMS OF SMALL ARCS IN SPACE OR TIME.

Arc.	0″	1″	2″	3″	4″	5″	6″	7″	8″	9″
0ʰ 20ᵐ 0ˢ	3.0792	3.0795	3.0799	3.0803	3.0806	3.0810	3.0813	3.0817	3.0821	3.0824
20 10	3.0828	3.0831	3.0835	3.0839	3.0842	3.0846	3.0849	3.0853	3.0856	3.0860
20 20	3.0864	3.0867	3.0871	3.0874	3.0878	3.0881	3.0885	3.0888	3.0892	3.0896
20 30	3.0899	3.0903	3.0906	3.0910	3.0913	3.0917	3.0920	3.0924	3.0927	3.0931
20 40	3.0934	3.0938	3.0941	3.0945	3.0948	3.0952	3.0955	3.0959	3.0962	3.0966
20 50	3.0969	3.0973	3.0976	3.0980	3.0983	3.0986	3.0990	3.0993	3.0997	3.1000
0 21 0	3.1004	3.1007	3.1011	3.1014	3.1017	3.1021	3.1024	3.1028	3.1031	3.1035
21 10	3.1038	3.1041	3.1045	3.1048	3.1052	3.1055	3.1059	3.1062	3.1065	3.1069
21 20	3.1072	3.1075	3.1079	3.1082	3.1086	3.1089	3.1092	3.1096	3.1099	3.1103
21 30	3.1106	3.1109	3.1113	3.1116	3.1119	3.1123	3.1126	3.1129	3.1133	3.1136
21 40	3.1139	3.1143	3.1146	3.1149	3.1153	3.1156	3.1159	3.1163	3.1166	3.1169
21 50	3.1173	3.1176	3.1179	3.1183	3.1186	3.1189	3.1193	3.1196	3.1199	3.1202
0 22 0	3.1206	3.1209	3.1212	3.1216	3.1219	3.1222	3.1225	3.1229	3.1232	3.1235
22 10	3.1239	3.1242	3.1245	3.1248	3.1252	3.1255	3.1258	3.1261	3.1265	3.1268
22 20	3.1271	3.1274	3.1278	3.1281	3.1284	3.1287	3.1290	3.1294	3.1297	3.1300
22 30	3.1303	3.1307	3.1310	3.1313	3.1316	3.1319	3.1323	3.1326	3.1329	3.1332
22 40	3.1335	3.1339	3.1342	3.1345	3.1348	3.1351	3.1355	3.1358	3.1361	3.1364
22 50	3.1367	3.1370	3.1374	3.1377	3.1380	3.1383	3.1386	3.1389	3.1392	3.1396
0 23 0	3.1399	3.1402	3.1405	3.1408	3.1411	3.1414	3.1418	3.1421	3.1424	3.1427
23 10	3.1430	3.1433	3.1436	3.1440	3.1443	3.1446	3.1449	3.1452	3.1455	3.1458
23 20	3.1461	3.1464	3.1467	3.1471	3.1474	3.1477	3.1480	3.1483	3.1486	3.1489
23 30	3.1492	3.1495	3.1498	3.1501	3.1504	3.1508	3.1511	3.1514	3.1517	3.1520
23 40	3.1523	3.1526	3.1529	3.1532	3.1535	3.1538	3.1541	3.1544	3.1547	3.1550
23 50	3.1553	3.1556	3.1559	3.1562	3.1565	3.1569	3.1572	3.1575	3.1578	3.1581
0 24 0	3.1584	3.1587	3.1590	3.1593	3.1596	3.1599	3.1602	3.1605	3.1608	3.1611
24 10	3.1614	3.1617	3.1620	3.1623	3.1626	3.1629	3.1632	3.1635	3.1638	3.1641
24 20	3.1644	3.1647	3.1649	3.1652	3.1655	3.1658	3.1661	3.1664	3.1667	3.1670
24 30	3.1673	3.1676	3.1679	3.1682	3.1685	3.1688	3.1691	3.1694	3.1697	3.1700
24 40	3.1703	3.1706	3.1708	3.1711	3.1714	3.1717	3.1720	3.1723	3.1726	3.1729
24 50	3.1732	3.1735	3.1738	3.1741	3.1744	3.1746	3.1749	3.1752	3.1755	3.1758
0 25 0	3.1761	3.1764	3.1767	3.1770	3.1772	3.1775	3.1778	3.1781	3.1784	3.1787
25 10	3.1790	3.1793	3.1796	3.1798	3.1801	3.1804	3.1807	3.1810	3.1813	3.1816
25 20	3.1818	3.1821	3.1824	3.1827	3.1830	3.1833	3.1836	3.1838	3.1841	3.1844
25 30	3.1847	3.1850	3.1853	3.1855	3.1858	3.1861	3.1864	3.1867	3.1870	3.1872
25 40	3.1875	3.1878	3.1881	3.1884	3.1886	3.1889	3.1892	3.1895	3.1898	3.1901
25 50	3.1903	3.1906	3.1909	3.1912	3.1915	3.1917	3.1920	3.1923	3.1926	3.1928
0 26 0	3.1931	3.1934	3.1937	3.1940	3.1942	3.1945	3.1948	3.1951	3.1953	3.1956
26 10	3.1959	3.1962	3.1965	3.1967	3.1970	3.1973	3.1976	3.1978	3.1981	3.1984
26 20	3.1987	3.1989	3.1992	3.1995	3.1998	3.2000	3.2003	3.2006	3.2009	3.2011
26 30	3.2014	3.2017	3.2019	3.2022	3.2025	3.2028	3.2030	3.2033	3.2036	3.2038
26 40	3.2041	3.2044	3.2047	3.2049	3.2052	3.2055	3.2057	3.2060	3.2063	3.2066
26 50	3.2068	3.2071	3.2074	3.2076	3.2079	3.2082	3.2084	3.2087	3.2090	3.2092
0 27 0	3.2095	3.2098	3.2101	3.2103	3.2106	3.2109	3.2111	3.2114	3.2117	3.2119
27 10	3.2122	3.2125	3.2127	3.2130	3.2133	3.2135	3.2138	3.2140	3.2143	3.2146
27 20	3.2148	3.2151	3.2154	3.2156	3.2159	3.2162	3.2164	3.2167	3.2170	3.2172
27 30	3.2175	3.2177	3.2180	3.2183	3.2185	3.2188	3.2191	3.2193	3.2196	3.2198
27 40	3.2201	3.2204	3.2206	3.2209	3.2212	3.2214	3.2217	3.2219	3.2222	3.2225
27 50	3.2227	3.2230	3.2232	3.2235	3.2238	3.2240	3.2243	3.2245	3.2248	3.2250
0 28 0	3.2253	3.2256	3.2258	3.2261	3.2263	3.2266	3.2269	3.2271	3.2274	3.2276
28 10	3.2279	3.2281	3.2284	3.2287	3.2289	3.2292	3.2294	3.2297	3.2299	3.2302
28 20	3.2304	3.2307	3.2310	3.2312	3.2315	3.2317	3.2320	3.2322	3.2325	3.2327
28 30	3.2330	3.2333	3.2335	3.2338	3.2340	3.2343	3.2345	3.2348	3.2350	3.2353
28 40	3.2355	3.2358	3.2360	3.2363	3.2365	3.2368	3.2370	3.2373	3.2375	3.2378
28 50	3.2380	3.2383	3.2385	3.2388	3.2390	3.2393	3.2395	3.2398	3.2400	3.2403
0 29 0	3.2405	3.2408	3.2410	3.2413	3.2415	3.2418	3.2420	3.2423	3.2425	3.2428
29 10	3.2430	3.2433	3.2435	3.2438	3.2440	3.2443	3.2445	3.2448	3.2450	3.2453
29 20	3.2455	3.2458	3.2460	3.2463	3.2465	3.2467	3.2470	3.2472	3.2475	3.2477
29 30	3.2480	3.2482	3.2485	3.2487	3.2490	3.2492	3.2494	3.2497	3.2499	3.2502
29 40	3.2504	3.2507	3.2509	3.2512	3.2514	3.2516	3.2519	3.2521	3.2524	3.2526
29 50	3.2529	3.2531	3.2533	3.2536	3.2538	3.2541	3.2543	3.2545	3.2548	3.2550

TABLE IX.

LOGARITHMS OF SMALL ARCS IN SPACE OR TIME.

Arc.	0″	1″	2″	3″	4″	5″	6″	7″	8″	9″
0ʰ·30ᵐ· 0ˢ	3.2553	3.2555	3.2558	3.2560	3.2562	3.2565	3.2567	3.2570	3.2572	3.2574
30 10	3.2577	3.2579	3.2582	3.2584	3.2586	3.2589	3.2591	3.2594	3.2596	3.2598
30 20	3.2601	3.2603	3.2605	3.2608	3.2610	3.2613	3.2615	3.2617	3.2620	3.2622
30 30	3.2625	3.2627	3.2629	3.2632	3.2634	3.2636	3.2639	3.2641	3.2643	3.2646
30 40	3.2648	3.2651	3.2653	3.2655	3.2658	3.2660	3.2662	3.2665	3.2667	3.2669
30 50	3.2672	3.2674	3.2676	3.2679	3.2681	3.2683	3.2686	3.2688	3.2690	3.2693
0 31 0	3.2695	3.2697	3.2700	3.2702	3.2704	3.2707	3.2709	3.2711	3.2714	3.2716
31 10	3.2718	3.2721	3.2723	3.2725	3.2728	3.2730	3.2732	3.2735	3.2737	3.2739
31 20	3.2742	3.2744	3.2746	3.2749	3.2751	3.2753	3.2755	3.2758	3.2760	3.2762
31 30	3.2765	3.2767	3.2769	3.2772	3.2774	3.2776	3.2778	3.2781	3.2783	3.2785
31 40	3.2788	3.2790	3.2792	3.2794	3.2797	3.2799	3.2801	3.2804	3.2806	3.2808
31 50	3.2810	3.2813	3.2815	3.2817	3.2819	3.2822	3.2824	3.2826	3.2828	3.2831
0 32 0	3.2833	3.2835	3.2838	3.2840	3.2842	3.2844	3.2847	3.2849	3.2851	3.2853
32 10	3.2856	3.2858	3.2860	3.2862	3.2865	3.2867	3.2869	3.2871	3.2874	3.2876
32 20	3.2878	3.2880	3.2882	3.2885	3.2887	3.2889	3.2891	3.2894	3.2896	3.2898
32 30	3.2900	3.2903	3.2905	3.2907	3.2909	3.2911	3.2914	3.2916	3.2918	3.2920
32 40	3.2923	3.2925	3.2927	3.2929	3.2931	3.2934	3.2936	3.2938	3.2940	3.2942
32 50	3.2945	3.2947	3.2949	3.2951	3.2953	3.2956	3.2958	3.2960	3.2962	3.2964
0 33 0	3.2967	3.2969	3.2971	3.2973	3.2975	3.2978	3.2980	3.2982	3.2984	3.2986
33 10	3.2989	3.2991	3.2993	3.2995	3.2997	3.2999	3.3002	3.3004	3.3006	3.3008
33 20	3.3010	3.3012	3.3015	3.3017	3.3019	3.3021	3.3023	3.3025	3.3028	3.3030
33 30	3.3032	3.3034	3.3036	3.3038	3.3041	3.3043	3.3045	3.3047	3.3049	3.3051
33 40	3.3054	3.3056	3.3058	3.3060	3.3062	3.3064	3.3066	3.3069	3.3071	3.3073
33 50	3.3075	3.3077	3.3079	3.3081	3.3084	3.3086	3.3088	3.3090	3.3092	3.3094
0 34 0	3.3096	3.3098	3.3101	3.3103	3.3105	3.3107	3.3109	3.3111	3.3113	3.3115
34 10	3.3118	3.3120	3.3122	3.3124	3.3126	3.3128	3.3130	3.3132	3.3134	3.3137
34 20	3.3139	3.3141	3.3143	3.3145	3.3147	3.3149	3.3151	3.3153	3.3156	3.3158
34 30	3.3160	3.3162	3.3164	3.3166	3.3168	3.3170	3.3172	3.3174	3.3176	3.3179
34 40	3.3181	3.3183	3.3185	3.3187	3.3189	3.3191	3.3193	3.3195	3.3197	3.3199
34 50	3.3201	3.3204	3.3206	3.3208	3.3210	3.3212	3.3214	3.3216	3.3218	3.3220
0 35 0	3.3222	3.3224	3.3226	3.3228	3.3230	3.3233	3.3235	3.3237	3.3239	3.3241
35 10	3.3243	3.3245	3.3247	3.3249	3.3251	3.3253	3.3255	3.3257	3.3259	3.3261
35 20	3.3263	3.3265	3.3267	3.3269	3.3272	3.3274	3.3276	3.3278	3.3280	3.3282
35 30	3.3284	3.3286	3.3288	3.3290	3.3292	3.3294	3.3296	3.3298	3.3300	3.3302
35 40	3.3304	3.3306	3.3308	3.3310	3.3312	3.3314	3.3316	3.3318	3.3320	3.3322
35 50	3.3324	3.3326	3.3328	3.3330	3.3332	3.3334	3.3336	3.3339	3.3341	3.3343
0 36 0	3.3345	3.3347	3.3349	3.3351	3.3353	3.3355	3.3357	3.3359	3.3361	3.3363
36 10	3.3365	3.3367	3.3369	3.3371	3.3373	3.3375	3.3377	3.3379	3.3381	3.3383
36 20	3.3385	3.3387	3.3389	3.3391	3.3393	3.3395	3.3397	3.3398	3.3400	3.3402
36 30	3.3404	3.3406	3.3408	3.3410	3.3412	3.3414	3.3416	3.3418	3.3420	3.3422
36 40	3.3424	3.3426	3.3428	3.3430	3.3432	3.3434	3.3436	3.3438	3.3440	3.3442
36 50	3.3444	3.3446	3.3448	3.3450	3.3452	3.3454	3.3456	3.3458	3.3460	3.3462
0 37 0	3.3464	3.3465	3.3467	3.3469	3.3471	3.3473	3.3475	3.3477	3.3479	3.3481
37 10	3.3483	3.3485	3.3487	3.3489	3.3491	3.3493	3.3495	3.3497	3.3499	3.3501
37 20	3.3502	3.3504	3.3506	3.3508	3.3510	3.3512	3.3514	3.3516	3.3518	3.3520
37 30	3.3522	3.3524	3.3526	3.3528	3.3530	3.3531	3.3533	3.3535	3.3537	3.3539
37 40	3.3541	3.3543	3.3545	3.3547	3.3549	3.3551	3.3553	3.3555	3.3556	3.3558
37 50	3.3560	3.3562	3.3564	3.3566	3.3568	3.3570	3.3572	3.3574	3.3576	3.3577
0 38 0	3.3579	3.3581	3.3583	3.3585	3.3587	3.3589	3.3591	3.3593	3.3595	3.3596
38 10	3.3598	3.3600	3.3602	3.3604	3.3606	3.3608	3.3610	3.3612	3.3614	3.3615
38 20	3.3617	3.3619	3.3621	3.3623	3.3625	3.3627	3.3629	3.3630	3.3632	3.3634
38 30	3.3636	3.3638	3.3640	3.3642	3.3644	3.3646	3.3647	3.3649	3.3651	3.3653
38 40	3.3655	3.3657	3.3659	3.3660	3.3662	3.3664	3.3666	3.3668	3.3670	3.3672
38 50	3.3674	3.3675	3.3677	3.3679	3.3681	3.3683	3.3685	3.3687	3.3688	3.3690
0 39 0	3.3692	3.3694	3.3696	3.3698	3.3700	3.3701	3.3703	3.3705	3.3707	3.3709
39 10	3.3711	3.3713	3.3714	3.3716	3.3718	3.3720	3.3722	3.3724	3.3725	3.3727
39 20	3.3729	3.3731	3.3733	3.3735	3.3736	3.3738	3.3740	3.3742	3.3744	3.3746
39 30	3.3747	3.3749	3.3751	3.3753	3.3755	3.3757	3.3758	3.3760	3.3762	3.3764
39 40	3.3766	3.3768	3.3769	3.3771	3.3773	3.3775	3.3777	3.3779	3.3780	3.3782
39 50	3.3784	3.3786	3.3788	3.3789	3.3791	3.3793	3.3795	3.3797	3.3798	3.3800

TABLE IX.

LOGARITHMS OF SMALL ARCS IN SPACE OR TIME.

Arc.	0″	1″	2″	3″	4″	5″	6″	7″	8″	9″
0ʰ 40ᵐ 0ˢ	3.3802	3.3804	3.3806	3.3808	3.3809	3.3811	3.3813	3.3815	3.3817	3.3818
40 10	3.3820	3.3822	3.3824	3.3826	3.3827	3.3829	3.3831	3.3833	3.3835	3.3836
40 20	3.3838	3.3840	3.3842	3.3844	3.3845	3.3847	3.3849	3.3851	3.3852	3.3854
40 30	3.3856	3.3858	3.3860	3.3861	3.3863	3.3865	3.3867	3.3869	3.3870	3.3872
40 40	3.3874	3.3876	3.3877	3.3879	3.3881	3.3883	3.3885	3.3886	3.3888	3.3890
40 50	3.3892	3.3893	3.3895	3.3897	3.3899	3.3901	3.3902	3.3904	3.3906	3.3908
41 0	3.3909	3.3911	3.3913	3.3915	3.3916	3.3918	3.3920	3.3922	3.3923	3.3925
41 10	3.3927	3.3929	3.3930	3.3932	3.3934	3.3936	3.3938	3.3939	3.3941	3.3943
41 20	3.3945	3.3946	3.3948	3.3950	3.3952	3.3953	3.3955	3.3957	3.3959	3.3960
41 30	3.3962	3.3964	3.3965	3.3967	3.3969	3.3971	3.3972	3.3974	3.3976	3.3978
41 40	3.3979	3.3981	3.3983	3.3985	3.3986	3.3988	3.3990	3.3992	3.3993	3.3995
41 50	3.3997	3.3998	3.4000	3.4002	3.4004	3.4005	3.4007	3.4009	3.4011	3.4012
0 42 0	3.4014	3.4016	3.4017	3.4019	3.4021	3.4023	3.4024	3.4026	3.4028	3.4029
42 10	3.4031	3.4033	3.4035	3.4036	3.4038	3.4040	3.4041	3.4043	3.4045	3.4047
42 20	3.4048	3.4050	3.4052	3.4053	3.4055	3.4057	3.4059	3.4060	3.4062	3.4064
42 30	3.4065	3.4067	3.4069	3.4071	3.4072	3.4074	3.4076	3.4077	3.4079	3.4081
42 40	3.4082	3.4084	3.4086	3.4087	3.4089	3.4091	3.4093	3.4094	3.4096	3.4098
42 50	3.4099	3.4101	3.4103	3.4104	3.4106	3.4108	3.4109	3.4111	3.4113	3.4115
0 43 0	3.4116	3.4118	3.4120	3.4121	3.4123	3.4125	3.4126	3.4128	3.4130	3.4131
43 10	3.4133	3.4135	3.4136	3.4138	3.4140	3.4141	3.4143	3.4145	3.4146	3.4148
43 20	3.4150	3.4151	3.4153	3.4155	3.4156	3.4158	3.4160	3.4161	3.4163	3.4165
43 30	3.4166	3.4168	3.4170	3.4171	3.4173	3.4175	3.4176	3.4178	3.4180	3.4181
43 40	3.4183	3.4185	3.4186	3.4188	3.4190	3.4191	3.4193	3.4195	3.4196	3.4198
43 50	3.4200	3.4201	3.4203	3.4205	3.4206	3.4208	3.4209	3.4211	3.4213	3.4214
0 44 0	3.4216	3.4218	3.4219	3.4221	3.4223	3.4224	3.4226	3.4228	3.4229	3.4231
44 10	3.4232	3.4234	3.4236	3.4237	3.4239	3.4241	3.4242	3.4244	3.4246	3.4247
44 20	3.4249	3.4250	3.4252	3.4254	3.4255	3.4257	3.4259	3.4260	3.4262	3.4263
44 30	3.4265	3.4267	3.4268	3.4270	3.4272	3.4273	3.4275	3.4276	3.4278	3.4280
44 40	3.4281	3.4283	3.4285	3.4286	3.4288	3.4289	3.4291	3.4293	3.4294	3.4296
44 50	3.4298	3.4299	3.4301	3.4302	3.4304	3.4306	3.4307	3.4309	3.4310	3.4312
0 45 0	3.4314	3.4315	3.4317	3.4318	3.4320	3.4322	3.4323	3.4325	3.4326	3.4328
45 10	3.4330	3.4331	3.4333	3.4334	3.4336	3.4338	3.4339	3.4341	3.4342	3.4344
45 20	3.4346	3.4347	3.4349	3.4350	3.4352	3.4354	3.4355	3.4357	3.4358	3.4360
45 30	3.4362	3.4363	3.4365	3.4366	3.4368	3.4370	3.4371	3.4373	3.4374	3.4376
45 40	3.4378	3.4379	3.4381	3.4382	3.4384	3.4385	3.4387	3.4389	3.4390	3.4392
45 50	3.4393	3.4395	3.4396	3.4398	3.4400	3.4401	3.4403	3.4404	3.4406	3.4408
0 46 0	3.4409	3.4411	3.4412	3.4414	3.4415	3.4417	3.4419	3.4420	3.4422	3.4423
46 10	3.4425	3.4426	3.4428	3.4429	3.4431	3.4433	3.4434	3.4436	3.4437	3.4439
46 20	3.4440	3.4442	3.4444	3.4445	3.4447	3.4448	3.4450	3.4451	3.4453	3.4454
46 30	3.4456	3.4458	3.4459	3.4461	3.4462	3.4464	3.4465	3.4467	3.4468	3.4470
46 40	3.4472	3.4473	3.4475	3.4476	3.4478	3.4479	3.4481	3.4482	3.4484	3.4486
46 50	3.4487	3.4489	3.4490	3.4492	3.4493	3.4495	3.4496	3.4498	3.4499	3.4501
0 47 0	3.4502	3.4504	3.4506	3.4507	3.4509	3.4510	3.4512	3.4513	3.4515	3.4516
47 10	3.4518	3.4519	3.4521	3.4522	3.4524	3.4526	3.4527	3.4529	3.4530	3.4532
47 20	3.4533	3.4535	3.4536	3.4538	3.4539	3.4541	3.4542	3.4544	3.4545	3.4547
47 30	3.4548	3.4550	3.4551	3.4553	3.4555	3.4556	3.4558	3.4559	3.4561	3.4562
47 40	3.4564	3.4565	3.4567	3.4568	3.4570	3.4571	3.4573	3.4574	3.4576	3.4577
47 50	3.4579	3.4580	3.4582	3.4583	3.4585	3.4586	3.4588	3.4589	3.4591	3.4592
0 48 0	3.4594	3.4595	3.4597	3.4598	3.4600	3.4601	3.4603	3.4604	3.4606	3.4607
48 10	3.4609	3.4610	3.4612	3.4613	3.4615	3.4616	3.4618	3.4619	3.4621	3.4622
48 20	3.4624	3.4625	3.4627	3.4628	3.4630	3.4631	3.4633	3.4634	3.4636	3.4637
48 30	3.4639	3.4640	3.4642	3.4643	3.4645	3.4646	3.4648	3.4649	3.4651	3.4652
48 40	3.4654	3.4655	3.4657	3.4658	3.4660	3.4661	3.4663	3.4664	3.4666	3.4667
48 50	3.4669	3.4670	3.4672	3.4673	3.4675	3.4676	3.4678	3.4679	3.4681	3.4682
0 49 0	3.4683	3.4685	3.4686	3.4688	3.4689	3.4691	3.4692	3.4694	3.4695	3.4697
49 10	3.4698	3.4700	3.4701	3.4703	3.4704	3.4706	3.4707	3.4709	3.4710	3.4711
49 20	3.4713	3.4714	3.4716	3.4717	3.4719	3.4720	3.4722	3.4723	3.4725	3.4726
49 30	3.4728	3.4729	3.4730	3.4732	3.4733	3.4735	3.4736	3.4738	3.4739	3.4741
49 40	3.4742	3.4744	3.4745	3.4747	3.4748	3.4749	3.4751	3.4752	3.4754	3.4755
49 50	3.4757	3.4758	3.4760	3.4761	3.4763	3.4764	3.4765	3.4767	3.4768	3.4770

TABLE IX.

LOGARITHMS OF SMALL ARCS IN SPACE OR TIME.

Arc.	0″	1″	2″	3″	4″	5″	6″	7″	8″	9′
0ʰ 50ᵐ 0ˢ	3.4771	3.4773	3.4774	3.4776	3.4777	3.4778	3.4780	3.4781	3.4783	3.4784
50 10	3.4786	3.4787	3.4789	3.4790	3.4791	3.4793	3.4794	3.4796	3.4797	3.4799
50 20	3.4800	3.4802	3.4803	3.4804	3.4806	3.4807	3.4809	3.4810	3.4812	3.4813
50 30	3.4814	3.4816	3.4817	3.4819	3.4820	3.4822	3.4823	3.4824	3.4826	3.4827
50 40	3.4829	3.4830	3.4832	3.4833	3.4834	3.4836	3.4837	3.4839	3.4840	3.4842
50 50	3.4843	3.4844	3.4846	3.4847	3.4849	3.4850	3.4852	3.4853	3.4854	3.4856
0 51 0	3.4857	3.4859	3.4860	3.4861	3.4863	3.4864	3.4866	3.4867	3.4869	3.4870
51 10	3.4871	3.4873	3.4874	3.4876	3.4877	3.4878	3.4880	3.4881	3.4883	3.4884
51 20	3.4886	3.4887	3.4888	3.4890	3.4891	3.4893	3.4894	3.4895	3.4897	3.4898
51 30	3.4900	3.4901	3.4902	3.4904	3.4905	3.4907	3.4908	3.4909	3.4911	3.4912
51 40	3.4914	3.4915	3.4916	3.4918	3.4919	3.4921	3.4922	3.4923	3.4925	3.4926
51 50	3.4928	3.4929	3.4930	3.4932	3.4933	3.4935	3.4936	3.4937	3.4939	3.4940
0 52 0	3.4942	3.4943	3.4944	3.4946	3.4947	3.4949	3.4950	3.4951	3.4953	3.4954
52 10	3.4955	3.4957	3.4958	3.4960	3.4961	3.4962	3.4964	3.4965	3.4967	3.4968
52 20	3.4969	3.4971	3.4972	3.4973	3.4975	3.4976	3.4978	3.4979	3.4980	3.4982
52 30	3.4983	3.4984	3.4986	3.4987	3.4989	3.4990	3.4991	3.4993	3.4994	3.4995
52 40	3.4997	3.4998	3.5000	3.5001	3.5002	3.5004	3.5005	3.5006	3.5008	3.5009
52 50	3.5011	3.5012	3.5013	3.5015	3.5016	3.5017	3.5019	3.5020	3.5022	3.5023
0 53 0	3.5024	3.5026	3.5027	3.5028	3.5030	3.5031	3.5032	3.5034	3.5035	3.5037
53 10	3.5038	3.5039	3.5041	3.5042	3.5043	3.5045	3.5046	3.5047	3.5049	3.5050
53 20	3.5051	3.5053	3.5054	3.5056	3.5057	3.5058	3.5060	3.5061	3.5062	3.5064
53 30	3.5065	3.5066	3.5068	3.5069	3.5070	3.5072	3.5073	3.5075	3.5076	3.5077
53 40	3.5079	3.5080	3.5081	3.5083	3.5084	3.5085	3.5087	3.5088	3.5089	3.5091
53 50	3.5092	3.5093	3.5095	3.5096	3.5097	3.5099	3.5100	3.5101	3.5103	3.5104
0 54 0	3.5105	3.5107	3.5108	3.5109	3.5111	3.5112	3.5113	3.5115	3.5116	3.5117
54 10	3.5119	3.5120	3.5122	3.5123	3.5124	3.5126	3.5127	3.5128	3.5130	3.5131
54 20	3.5132	3.5134	3.5135	3.5136	3.5138	3.5139	3.5140	3.5141	3.5143	3.5144
54 30	3.5145	3.5147	3.5148	3.5149	3.5151	3.5152	3.5153	3.5155	3.5156	3.5157
54 40	3.5159	3.5160	3.5161	3.5163	3.5164	3.5165	3.5167	3.5168	3.5169	3.5171
54 50	3.5172	3.5173	3.5175	3.5176	3.5177	3.5179	3.5180	3.5181	3.5183	3.5184
0 55 0	3.5185	3.5186	3.5188	3.5189	3.5190	3.5192	3.5193	3.5194	3.5196	3.5197
55 10	3.5198	3.5200	3.5201	3.5202	3.5204	3.5205	3.5206	3.5207	3.5209	3.5210
55 20	3.5211	3.5213	3.5214	3.5215	3.5217	3.5218	3.5219	3.5221	3.5222	3.5223
55 30	3.5224	3.5226	3.5227	3.5228	3.5230	3.5231	3.5232	3.5234	3.5235	3.5236
55 40	3.5237	3.5239	3.5240	3.5241	3.5243	3.5244	3.5245	3.5247	3.5248	3.5249
55 50	3.5250	3.5252	3.5253	3.5254	3.5256	3.5257	3.5258	3.5260	3.5261	3.5262
0 56 0	3.5263	3.5265	3.5266	3.5267	3.5269	3.5270	3.5271	3.5272	3.5274	3.5275
56 10	3.5276	3.5278	3.5279	3.5280	3.5281	3.5283	3.5284	3.5285	3.5287	3.5288
56 20	3.5289	3.5290	3.5292	3.5293	3.5294	3.5296	3.5297	3.5298	3.5299	3.5301
56 30	3.5302	3.5303	3.5305	3.5306	3.5307	3.5308	3.5310	3.5311	3.5312	3.5314
56 40	3.5315	3.5316	3.5317	3.5319	3.5320	3.5321	3.5322	3.5324	3.5325	3.5326
56 50	3.5328	3.5329	3.5330	3.5331	3.5333	3.5334	3.5335	3.5336	3.5338	3.5339
0 57 0	3.5340	3.5342	3.5343	3.5344	3.5345	3.5347	3.5348	3.5349	3.5350	3.5352
57 10	3.5353	3.5354	3.5355	3.5357	3.5358	3.5359	3.5361	3.5362	3.5363	3.5364
57 20	3.5366	3.5367	3.5368	3.5369	3.5371	3.5372	3.5373	3.5374	3.5376	3.5377
57 30	3.5378	3.5379	3.5391	3.5382	3.5383	3.5384	3.5386	3.5387	3.5388	3.5390
57 40	3.5391	3.5392	3.5393	3.5395	3.5396	3.5397	3.5398	3.5400	3.5401	3.5402
57 50	3.5403	3.5405	3.5406	3.5407	3.5408	3.5410	3.5411	3.5412	3.5413	3.5415
0 58 0	3.5416	3.5417	3.5418	3.5420	3.5421	3.5422	3.5423	3.5425	3.5426	3.5427
58 10	3.5428	3.5429	3.5431	3.5432	3.5433	3.5434	3.5436	3.5437	3.5438	3.5439
58 20	3.5441	3.5442	3.5443	3.5444	3.5446	3.5447	3.5448	3.5449	3.5451	3.5452
58 30	3.5453	3.5454	3.5456	3.5457	3.5458	3.5459	3.5460	3.5462	3.5463	3.5464
58 40	3.5465	3.5467	3.5468	3.5469	3.5470	3.5472	3.5473	3.5474	3.5475	3.5477
58 50	3.5478	3.5479	3.5480	3.5481	3.5483	3.5484	3.5485	3.5486	3.5488	3.5489
0 59 0	3.5490	3.5491	3.5492	3.5494	3.5495	3.5496	3.5497	3.5499	3.5500	3.5501
59 10	3.5502	3.5504	3.5505	3.5506	3.5507	3.5508	3.5510	3.5511	3.5512	3.5513
59 20	3.5514	3.5516	3.5517	3.5518	3.5519	3.5521	3.5522	3.5523	3.5524	3.5525
59 30	3.5527	3.5528	3.5529	3.5530	3.5532	3.5533	3.5534	3.5535	3.5536	3.5538
59 40	3.5539	3.5540	3.5541	3.5542	3.5544	3.5545	3.5546	3.5547	3.5549	3.5550
59 50	3.5551	3.5552	3.5553	3.5555	3.5556	3.5557	3.5558	3.5559	3.5561	3.5562

TABLE IX.

LOGARITHMS OF SMALL ARCS IN SPACE OR TIME.

Arc.	$0''$	$1''$	$2''$	$3''$	$4''$	$5''$	$6''$	$7''$	$8''$	$9''$
h. 0ᵐ 0ˢ	3.5563	3.5564	3.5565	3.5567	3.5568	3.5569	3.5570	3.5571	3.5573	3.5574
0 10	3.5575	3.5576	3.5577	3.5579	3.5580	3.5581	3.5582	3.5583	3.5585	3.5586
0 20	3.5587	3.5588	3.5589	3.5591	3.5592	3.5593	3.5594	3.5595	3.5597	3.5598
0 30	3.5599	3.5600	3.5601	3.5603	3.5604	3.5605	3.5606	3.5607	3.5609	3.5610
0 40	3.5611	3.5612	3.5613	3.5615	3.5616	3.5617	3.5618	3.5619	3.5621	3.5622
0 50	3.5623	3.5624	3.5625	3.5626	3.5628	3.5629	3.5630	3.5631	3.5632	3.5634
1 1 0	3.5635	3.5636	3.5637	3.5638	3.5640	3.5641	3.5642	3.5643	3.5644	3.5645
1 10	3.5647	3.5648	3.5649	3.5650	3.5651	3.5653	3.5654	3.5655	3.5656	3.5657
1 20	3.5658	3.5660	3.5661	3.5662	3.5663	3.5664	3.5666	3.5667	3.5668	3.5669
1 30	3.5670	3.5671	3.5673	3.5674	3.5675	3.5676	3.5677	3.5678	3.5680	3.5681
1 40	3.5682	3.5683	3.5684	3.5686	3.5687	3.5688	3.5689	3.5690	3.5691	3.5693
1 50	3.5694	3.5695	3.5696	3.5697	3.5699	3.5700	3.5701	3.5702	3.5703	3.5704
1 2 0	3.5705	3.5707	3.5708	3.5709	3.5710	3.5711	3.5712	3.5714	3.5715	3.5716
2 10	3.5717	3.5718	3.5719	3.5721	3.5722	3.5723	3.5724	3.5725	3.5726	3.5728
2 20	3.5729	3.5730	3.5731	3.5732	3.5733	3.5735	3.5736	3.5737	3.5738	3.5739
2 30	3.5740	3.5741	3.5742	3.5744	3.5745	3.5746	3.5747	3.5748	3.5750	3.5751
2 40	3.5752	3.5753	3.5754	3.5755	3.5756	3.5758	3.5759	3.5760	3.5761	3.5762
2 50	3.5763	3.5765	3.5766	3.5767	3.5768	3.5769	3.5770	3.5771	3.5773	3.5774
1 3 0	3.5775	3.5776	3.5777	3.5778	3.5780	3.5781	3.5782	3.5783	3.5784	3.5785
3 10	3.5786	3.5788	3.5789	3.5790	3.5791	3.5792	3.5793	3.5794	3.5796	3.5797
3 20	3.5798	3.5799	3.5800	3.5801	3.5802	3.5804	3.5805	3.5806	3.5807	3.5808
3 30	3.5809	3.5810	3.5812	3.5813	3.5814	3.5815	3.5816	3.5817	3.5818	3.5819
3 40	3.5821	3.5822	3.5823	3.5824	3.5825	3.5826	3.5827	3.5829	3.5830	3.5831
3 50	3.5832	3.5833	3.5834	3.5835	3.5837	3.5838	3.5839	3.5840	3.5841	3.5842
1 4 0	3.5843	3.5844	3.5846	3.5847	3.5848	3.5849	3.5850	3.5851	3.5852	3.5853
4 10	3.5855	3.5856	3.5857	3.5858	3.5859	3.5860	3.5861	3.5862	3.5864	3.5865
4 20	3.5866	3.5867	3.5868	3.5869	3.5870	3.5871	3.5873	3.5874	3.5875	3.5876
4 30	3.5877	3.5878	3.5879	3.5880	3.5882	3.5883	3.5884	3.5885	3.5886	3.5887
4 40	3.5888	3.5889	3.5891	3.5892	3.5893	3.5894	3.5895	3.5896	3.5897	3.5898
4 50	3.5899	3.5901	3.5902	3.5903	3.5904	3.5905	3.5906	3.5907	3.5908	3.5910
1 5 0	3.5911	3.5912	3.5913	3.5914	3.5915	3.5916	3.5917	3.5918	3.5920	3.5921
5 10	3.5922	3.5923	3.5924	3.5925	3.5926	3.5927	3.5928	3.5930	3.5931	3.5932
5 20	3.5933	3.5934	3.5935	3.5936	3.5937	3.5938	3.5940	3.5941	3.5942	3.5943
5 30	3.5944	3.5945	3.5946	3.5947	3.5948	3.5949	3.5951	3.5952	3.5953	3.5954
5 40	3.5955	3.5956	3.5957	3.5958	3.5959	3.5960	3.5962	3.5963	3.5964	3.5965
5 50	3.5966	3.5967	3.5968	3.5969	3.5970	3.5971	3.5973	3.5974	3.5975	3.5976
1 6 0	3.5977	3.5978	3.5979	3.5980	3.5981	3.5982	3.5984	3.5985	3.5986	3.5987
6 10	3.5988	3.5989	3.5990	3.5991	3.5992	3.5993	3.5994	3.5996	3.5997	3.5998
6 20	3.5999	3.6000	3.6001	3.6002	3.6003	3.6004	3.6005	3.6006	3.6008	3.6009
6 30	3.6010	3.6011	3.6012	3.6013	3.6014	3.6015	3.6016	3.6017	3.6018	3.6020
6 40	3.6021	3.6022	3.6023	3.6024	3.6025	3.6026	3.6027	3.6028	3.6029	3.6030
6 50	3.6031	3.6033	3.6034	3.6035	3.6036	3.6037	3.6038	3.6039	3.6040	3.6041
1 7 0	3.6042	3.6043	3.6044	3.6046	3.6047	3.6048	3.6049	3.6050	3.6051	3.6052
7 10	3.6053	3.6054	3.6055	3.6056	3.6057	3.6058	3.6060	3.6061	3.6062	3.6063
7 20	3.6064	3.6065	3.6066	3.6067	3.6068	3.6069	3.6070	3.6071	3.6072	3.6073
7 30	3.6075	3.6076	3.6077	3.6078	3.6079	3.6080	3.6081	3.6082	3.6083	3.6084
7 40	3.6085	3.6086	3.6087	3.6088	3.6090	3.6091	3.6092	3.6093	3.6094	3.6095
7 50	3.6096	3.6097	3.6098	3.6099	3.6100	3.6101	3.6102	3.6103	3.6104	3.6106
1 8 0	3.6107	3.6108	3.6109	3.6110	3.6111	3.6112	3.6113	3.6114	3.6115	3.6116
8 10	3.6117	3.6118	3.6119	3.6120	3.6121	3.6123	3.6124	3.6125	3.6126	3.6127
8 20	3.6128	3.6129	3.6130	3.6131	3.6132	3.6133	3.6134	3.6135	3.6136	3.6137
8 30	3.6138	3.6139	3.6141	3.6142	3.6143	3.6144	3.6145	3.6146	3.6147	3.6148
8 40	3.6149	3.6150	3.6151	3.6152	3.6153	3.6154	3.6155	3.6156	3.6157	3.6158
8 50	3.6160	3.6161	3.6162	3.6163	3.6164	3.6165	3.6166	3.6167	3.6168	3.6169
1 9 0	3.6170	3.6171	3.6172	3.6173	3.6174	3.6175	3.6176	3.6177	3.6178	3.6179
9 10	3.6180	3.6182	3.6183	3.6184	3.6185	3.6186	3.6187	3.6188	3.6189	3.6190
9 20	3.6191	3.6192	3.6193	3.6194	3.6195	3.6196	3.6197	3.6198	3.6199	3.6200
9 30	3.6201	3.6202	3.6203	3.6204	3.6206	3.6207	3.6208	3.6209	3.6210	3.6211
9 40	3.6212	3.6213	3.6214	3.6215	3.6216	3.6217	3.6218	3.6219	3.6220	3.6221
9 50	3.6222	3.6223	3.6224	3.6225	3.6226	3.6227	3.6228	3.6229	3.6230	3.6231

TABLE IX.

LOGARITHMS OF SMALL ARCS IN SPACE OR TIME.										
Arc.	0″	1″	2″	3″	4″	5″	6″	7″	8″	9″
1ʰ.10ᵐ. 0ˢ.	3.6232	3.6234	3.6235	3.6236	3.6237	3.6238	3.6239	3.6240	3.6241	3.6242
10 10	3.6243	3.6244	3.6245	3.6246	3.6247	3.6248	3.6249	3.6250	3.6251	3.6252
10 20	3.6253	3.6254	3.6255	3.6256	3.6257	3.6258	3.6259	3 6260	3.6261	3.6262
10 30	3.6263	3.6264	3.6265	3.6266	3.6268	3.6269	3.6270	3.6271	3.0272	3.6273
10 40	3.6274	3.6275	3.6276	3.6277	3.6278	3.6279	3.6280	3.6281	3.6282	3.6283
10 50	3.6284	3.6285	3.6286	3.6287	3.6288	3.6289	3.6290	3.6291	3.6292	3.6293
1 11 0	3.6294	3.6295	3.6296	3.6297	3.9298	3.6299	3.6300	3.6301	3.6302	3.6303
11 10	3.6304	3.6305	3.6306	3.6307	3.6308	3.6309	3.6310	3.6311	3 6312	3.6313
11 20	3.6314	3.6315	3.6316	3.6317	3.6318	3.6320	3.6321	3.6322	3.6323	3.6324
11 30	3.6325	3.6326	3.6327	3.6328	3.6329	3.6330	3.6331	3.6332	3.6333	3.6334
11 40	3.6335	3.6336	3.6337	3.6338	3.6339	3.6340	3.6341	3.6342	3.6343	3.6344
11 50	3.6345	3.6346	3.6347	3.6348	3.6349	3.6350	3.6351	3.6352	3.6353	3.6354
1 12 0	3.6355	3.6356	3.6357	3.6358	3.6359	3.6360	3.6361	3.6362	3.6363	3.6364
12 10	3.6365	3.6366	3.6367	3.6368	3.6369	3.6370	3.6371	3.6372	3.6373	3.6374
12 20	3.6375	3.6376	3.6377	3.6378	3.6379	3.6380	3.6381	3.6382	3.6383	3.6384
12 30	3.6385	3.6386	3.6387	3.6388	3.6389	3.6390	3.6391	3.6392	3.6393	3.6394
12 40	3.6395	3.6396	3.6397	3.6398	3.6399	3.6400	3.6401	3.6402	3.6403	3 6404
12 50	3.6405	3.6406	3.6407	3.6408	3.6409	3.6410	3.6411	3.6412	3.6413	3.6414
1 13 0	3.6415	3.6416	3.6417	3.6418	3.6419	3.6420	3.6421	3.6422	3.6423	3.6424
13 10	3.6425	3.6426	3.6427	3.6428	3.6429	3.6430	3.6431	3.6432	3.6433	3.6434
13 20	3.6435	3.6436	3.6437	3.6437	3.6438	3.6439	3.6440	3.6441	3.6442	3.6443
13 30	3.6444	3.6445	3.6446	3.6447	3.6448	3.6449	3.6450	3.6451	3.6452	3.6453
13 40	3.6454	3.6455	3.6456	3.6457	3.6458	3.6459	3.6460	3.6461	3.6462	3.6463
13 50	3.6464	3.6465	3.6466	3.6467	3.6468	3.6469	3.6470	3.6471	3.6472	3.6473
1 14 0	3.6474	3.6475	3.6476	3.6477	3.6478	3.6479	3.6480	3.6481	3.6482	3.6483
14 10	3.6484	3.6485	3.6486	3.6487	3.6488	3.6488	3.6489	3.6490	3.6491	3.6492
14 20	3.6493	3.6494	3.6495	3.6496	3.6497	3.6498	3.6499	3.6500	3.6501	3.6502
14 30	3.6503	3.6504	3.6505	3.6506	3.6507	3.6508	3.6509	3.6510	3.6511	3.6512
14 40	3.6513	3.6514	3.6515	3.6516	3.6517	3.6518	3.6519	3.6520	3.6521	3.6521
14 50	3.6522	3.6523	3.6524	3.6525	3.6526	3.6527	3.6528	3.6529	3.6530	3.6531
1 15 0	3.6532	3.6533	3.6534	3.6535	3.6536	3.6537	3.6538	3.6539	3.6540	3.6541
15 10	3.6542	3.6543	3.6544	3.6545	3.6546	3.6547	3.6548	3.6549	3.6549	3.6550
15 20	3.6551	3.6552	3.6553	3.6554	3.6555	3.6556	3.6557	3.6558	3.6559	3.6560
15 30	3.6561	3.6562	3.6563	3.6564	3.6565	3.6566	3.6567	3.6568	3.6569	3.6570
15 40	3.6571	3.6572	3.6572	3.6573	3.6574	3.6575	3.6576	3.6577	3.6578	3.6579
15 50	3.6580	3.6581	3.6582	3.6583	3.6584	3.6585	3.6586	3.6587	3 6588	3.6589
1 16 0	3.6590	3.6591	3.6592	3.6593	3.6593	3.6594	3.6595	3.6596	3.6597	3.6598
16 10	3.6599	3.6600	3.6601	3.6602	3.6603	3.6604	3.6605	3.6606	3.6607	3.6608
16 20	3.6609	3.6610	3.6611	3.6611	3.6612	3.6613	3.6614	3.6615	3.6616	3.6617
16 30	3.6618	3.6619	3.6620	3.6621	3.6622	3.6623	3.6624	3.6625	3.6626	3.6627
16 40	3.6628	3.6629	3.6629	3.6630	3.6631	3 6632	3.6633	3.6634	3.6635	3.6636
16 50	3.6637	3.6638	3.6639	3.6640	3.6641	3.6642	3.6643	3.6644	3.6645	3.6645
1 17 0	3.6646	3.6647	3.6648	3.6649	3.6650	3.6651	3.6652	3.6653	3.6654	3.6655
17 10	3.6656	3.6657	3.6658	3.6659	3.6660	3.6660	3.6661	3.6662	3.6663	3.6664
17 20	3.6665	3.6666	3.6667	3.6668	3.6669	3.6670	3.6671	3.6672	3.6673	3.6674
17 30	3.6675	3.6675	3.6676	3.6677	3.6678	3.6679	3.6680	3.6681	3.6682	3.6683
17 40	3.6684	3.6685	3.6686	3.6687	3.6688	3.6689	3.6689	3.6690	3.6691	3.6692
17 50	3.6693	3.6694	3.6695	3.6696	3.6697	3.6698	3.6699	3.6700	3.6701	3.6702
1 18 0	3.6702	3.6703	3.6704	3.6705	3.6706	3.6707	3.6708	3.6709	3.6710	3.6711
18 10	3.6712	3.6713	3.6714	3.6715	3.6715	3.6716	3.6717	3.6718	3.6719	3.6720
18 20	3.6721	3.6722	3.6723	3.6724	3.6725	3.6726	3.6727	3.6727	3.6728	3.6729
18 30	3.6730	3.6731	3.6732	3.6733	3.6734	3.6735	3.6736	3.6737	3.6738	3.6738
18 40	3.6739	3.6740	3.6741	3.6742	3 6743	3.6744	3.6745	3.6746	3.6747	3.6748
18 50	3.6749	3.6750	3.6750	3.6751	3.6752	3.6753	3.6754	3.6755	3.6756	3.6757
1 19 0	3.6758	3.6759	3.6760	3.6761	3.6761	3.6762	3.6763	3.6764	3.6765	3.6766
19 10	3.6767	3.6768	3.6769	3.6770	3.6771	3.6772	3.6772	3.6773	3.6774	3.6775
19 20	3.6776	3.6777	3.6778	3.6779	3.6780	3.6781	3.6782	3.6782	3.6783	3.6784
19 30	3.6785	3.6786	3.6787	3.6788	3.6789	3.6790	3.6791	3.6792	3.6792	3.6793
19 40	3.6794	3.6795	3 6796	3.6797	3.6798	3.6799	3.6800	3.6801	3.6802	3.6802
19 50	3.6803	3.6804	3.6805	3.6806	3.6807	3.6808	3.6809	3.6810	3.6811	3.6812

TABLE IX.

LOGARITHMS OF SMALL ARCS IN SPACE OR TIME.

Arc.	0″	1″	2″	3″	4″	5″	6″	7″	8″	9″
1ʰ 20ᵐ 0ˢ	3.6812	3.6813	3.6814	3.6815	3.6816	3.6817	3.6818	3.6819	3.6820	3.6821
20 10	3.6821	3.6822	3.6823	3.6824	3.6825	3.6826	3.6827	3.6828	3.6829	3.6830
20 20	3.6830	3.6831	3.6832	3.6833	3.6834	3.6835	3.6836	3.6837	3.6838	3.6839
20 30	3.6839	3.6840	3.6841	3.6842	3.6843	3.6844	3.6845	3.6846	3.6847	3.6848
20 40	3.6848	3.6849	3.6850	3.6851	3.6852	3.6853	3.6854	3.6855	3.6856	3.6857
20 50	3.6857	3.6858	3.6859	3.6860	3.6861	3.6862	3.6863	3.6864	3.6865	3.6865
1 21 0	3.6866	3.6867	3.6868	3.6869	3.6870	3.6871	3.6872	3.6873	3.6874	3.6874
21 10	3.6875	3.6876	3.6877	3.6878	3.6879	3.6880	3.6881	3.6882	3.6882	3.6883
21 20	3.6884	3.6885	3.6886	3.6887	3.6888	3.6889	3.6890	3.6890	3.6891	3.6892
21 30	3.6893	3.6894	3.6895	3.6896	3.6897	3.6898	3.6898	3.6899	3.6900	3.6901
21 40	3.6902	3.6903	3.6904	3.6905	3.6906	3.6906	3.6907	3.6908	3.6909	3.6910
21 50	3.6911	3.6912	3.6913	3.6913	3.6914	3.6915	3.6916	3.6917	3.6918	3.6919
1 22 0	3.6920	3.6921	3.6921	3.6922	3.6923	3.6924	3.6925	3.6926	3.6927	3.6928
22 10	3.6928	3.6929	3.6930	3.6931	3.6932	3.6933	3.6934	3.6935	3.6936	3.6936
22 20	3.6937	3.6938	3.6939	3.6940	3.6941	3.6942	3.6943	3.6943	3.6944	3.6945
22 30	3.6946	3.6947	3.6948	3.6949	3.6950	3.6950	3.6951	3.6952	3.6953	3.6954
22 40	3.6955	3.6956	3.6957	3.6957	3.6958	3.6959	3.6960	3.6961	3.6962	3.6963
22 50	3.6964	3.6964	3.6965	3.6966	3.6967	3.6968	3.6969	3.6970	3.6971	3.6971
1 23 0	3.6972	3.6973	3.6974	3.6975	3.6976	3.6977	3.6978	3.6978	3.6979	3.6980
23 10	3.6981	3.6982	3.6983	3.6984	3.6984	3.6985	3.6986	3.6987	3.6988	3.6989
23 20	3.6990	3.6991	3.6991	3.6992	3.6993	3.6994	3.6995	3.6996	3.6997	3.6998
23 30	3.6998	3.6999	3.7000	3.7001	3.7002	3.7003	3.7004	3.7004	3.7005	3.7006
23 40	3.7007	3.7008	3.7009	3.7010	3.7010	3.7011	3.7012	3.7013	3.7014	3.7015
23 50	3.7016	3.7017	3.7017	3.7018	3.7019	3.7020	3.7021	3.7022	3.7023	3.7023
1 24 0	3.7024	3.7025	3.7026	3.7027	3.7028	3.7029	3.7029	3.7030	3.7031	3.7032
24 10	3.7033	3.7034	3.7035	3.7035	3.7036	3.7037	3.7038	3.7039	3.7040	3.7041
24 20	3.7042	3.7042	3.7043	3.7044	3.7045	3.7046	3.7047	3.7048	3.7048	3.7049
24 30	3.7050	3.7051	3.7052	3.7053	3.7054	3.7054	3.7055	3.7056	3.7057	3.7058
24 40	3.7059	3.7060	3.7060	3.7061	3.7062	3.7063	3.7064	3.7065	3.7065	3.7066
24 50	3.7067	3.7068	3.7069	3.7070	3.7071	3.7071	3.7072	3.7073	3.7074	3.7075
1 25 0	3.7076	3.7077	3.7077	3.7078	3.7079	3.7080	3.7081	3.7082	3.7083	3.7083
25 10	3.7084	3.7085	3.7086	3.7087	3.7088	3.7088	3.7089	3.7090	3.7091	3.7092
25 20	3.7093	3.7094	3.7094	3.7095	3.7096	3.7097	3.7098	3.7099	3.7099	3.7100
25 30	3.7101	3.7102	3.7103	3.7104	3.7105	3.7105	3.7106	3.7107	3.7108	3.7109
25 40	3.7110	3.7110	3.7111	3.7112	3.7113	3.7114	3.7115	3.7116	3.7116	3.7117
25 50	3.7118	3.7119	3.7120	3.7121	3.7121	3.7122	3.7123	3.7124	3.7125	3.7126
1 26 0	3.7126	3.7127	3.7128	3.7129	3.7130	3.7131	3.7132	3.7132	3.7133	3.7134
26 10	3.7135	3.7136	3.7137	3.7137	3.7138	3.7139	3.7140	3.7141	3.7142	3.7142
26 20	3.7143	3.7144	3.7145	3.7146	3.7147	3.7147	3.7148	3.7149	3.7150	3.7151
26 30	3.7152	3.7153	3.7153	3.7154	3.7155	3.7156	3.7157	3.7158	3.7159	3.7159
26 40	3.7160	3.7161	3.7162	3.7163	3.7163	3.7164	3.7165	3.7166	3.7167	3.7168
26 50	3.7168	3.7169	3.7170	3.7171	3.7172	3.7173	3.7173	3.7174	3.7175	3.7176
1 27 0	3.7177	3.7178	3.7178	3.7179	3.7180	3.7181	3.7182	3.7183	3.7183	3.7184
27 10	3.7185	3.7186	3.7187	3.7188	3.7188	3.7189	3.7190	3.7191	3.7192	3.7192
27 20	3.7193	3.7194	3.7195	3.7196	3.7197	3.7197	3.7198	3.7199	3.7200	3.7201
27 30	3.7202	3.7202	3.7203	3.7204	3.7205	3.7206	3.7207	3.7207	3.7208	3.7209
27 40	3.7210	3.7211	3.7212	3.7212	3.7213	3.7214	3.7215	3.7216	3.7216	3.7217
27 50	3.7218	3.7219	3.7220	3.7221	3.7221	3.7222	3.7223	3.7224	3.7225	3.7226
1 28 0	3.7226	3.7227	3.7228	3.7229	3.7230	3.7230	3.7231	3.7232	3.7233	3.7234
28 10	3.7235	3.7235	3.7236	3.7237	3.7238	3.7239	3.7239	3.7240	3.7241	3.7242
28 20	3.7243	3.7244	3.7244	3.7245	3.7246	3.7247	3.7248	3.7248	3.7249	3.7250
28 30	3.7251	3.7252	3.7253	3.7253	3.7254	3.7255	3.7256	3.7257	3.7257	3.7258
28 40	3.7259	3.7260	3.7261	3.7262	3.7262	3.7263	3.7264	3.7265	3.7266	3.7266
28 50	3.7267	3.7268	3.7269	3.7270	3.7271	3.7271	3.7272	3.7273	3.7274	3.7275
1 29 0	3.7275	3.7276	3.7277	3.7278	3.7279	3.7279	3.7280	3.7281	3.7282	3.7283
29 10	3.7284	3.7284	3.7285	3.7286	3.7287	3.7288	3.7288	3.7289	3.7290	3.7291
29 20	3.7292	3.7292	3.7293	3.7294	3.7295	3.7296	3.7297	3.7297	3.7298	3.7299
29 30	3.7300	3.7301	3.7301	3.7302	3.7303	3.7304	3.7305	3.7305	3.7306	3.7307
29 40	3.7308	3.7309	3.7309	3.7310	3.7311	3.7312	3.7313	3.7313	3.7314	3.7315
29 50	3.7316	3.7317	3.7317	3.7318	3.7319	3.7320	3.7321	3.7322	3.7322	3.7323

TABLE X.

LOGARITHMS OF SMALL ARCS IN SPACE OR TIME.

Arc.	0″	1″	2″	3″	4″	5″	6″	7″	8″	9″
1ʰ·30ᵐ· 0ˢ·	3.7324	3.7325	3.7326	3.7326	3.7327	3.7328	3.7329	3 7330	3.7330	3.7331
30 10	3.7332	3.7333	3.7334	3.7334	3.7335	3.7336	3.7337	3.7338	3.7338	3.7339
30 20	3.7340	3.7341	3.7342	3.7342	3.7343	3.7344	3.7345	3.7346	3.7346	3.7347
30 30	3.7348	3.7349	3.7350	3.7350	3.7351	3.7352	3.7353	3.7354	3.7354	3.7355
30 40	3.7356	3.7357	3.7358	3.7358	3.7359	3.7360	3.7361	3.7362	3.7362	3.7363
30 50	3.7364	3.7365	3.7366	3.7366	3.7367	3.7368	3.7369	3.7370	3.7370	3.7371
1 31 0	3.7372	3.7373	3.7374	3.7374	3.7375	3.7376	3.7377	3.7377	3.7378	3.7379
31 10	3.7380	3.7381	3.7381	3.7382	3.7383	3.7384	3.7385	3.7385	3.7386	3.7387
31 20	3.7388	3.7389	3.7389	3.7390	3 7391	3.7392	3.7393	3.7393	3.7394	3.7395
31 30	3.7396	3.7397	3.7397	3.7398	3.7399	3.7400	3.7400	3.7401	3.7402	3.7403
31 40	3.7404	3.7404	3.7405	3.7406	3.7407	3.7408	3.7408	3.7409	3.7410	3.7411
31 50	3.7412	3.7412	3.7413	3.7414	3.7415	3.7415	3.7416	3.7417	3.7418	3.7419
1 32 0	3.7419	3.7420	3.7421	3.7422	3.7423	3.7423	3.7424	3.7425	3.7426	3.7426
32 10	3.7427	3.7428	3.7429	3.7430	3.7430	3.7431	3.7432	3.7433	3.7434	3.7434
32 20	3.7435	3.7436	3.7437	3.7437	3.7438	3.7439	3.7440	3.7441	3.7441	3.7442
32 30	3.7443	3.7444	3.7444	3.7445	3.7446	3.7447	3.7448	3.7448	3.7449	3.7450
32 40	3.7451	3.7452	3.7452	3.7453	3.7454	3.7455	3.7455	3.7456	3.7457	3.7458
32 50	3.7459	3.7459	3.7460	3.7461	3.7462	3.7462	3.7463	3.7464	3.7465	3.7466
1 33 0	3.7466	3.7467	3.7468	3.7469	3.7469	3.7470	3.7471	3.7472	3.7473	3.7473
33 10	3.7474	3.7475	3.7476	3.7476	3.7477	3.7478	3.7479	3.7480	3.7480	3.7481
33 20	3.7482	3.7483	3.7483	3.7484	3.7485	3.7486	3.7487	3.7487	3.7488	3.7489
33 30	3.7490	3.7490	3.7491	3.7492	3.7493	3.7493	3.7494	3.7495	3.7496	3.7497
33 40	3.7497	3.7498	3.7499	3.7500	3.7500	3.7501	3.7502	3.7503	3.7504	3.7504
33 50	3.7505	3.7506	3.7507	3.7507	3.7508	3.7509	3.7510	3.7510	3.7511	3.7512
1 34 0	3.7513	3.7514	3.7514	3.7515	3.7516	3.7517	3.7517	3.7518	3.7519	3.7520
34 10	3.7520	3.7521	3.7522	3.7523	3.7524	3.7524	3.7525	3.7526	3.7527	3.7527
34 20	3.7528	3.7529	3.7530	3.7530	3.7531	3.7532	3.7533	3.7534	3.7534	3.7535
34 30	3.7536	3.7537	3.7537	3.7538	3.7539	3.7540	3.7540	3.7541	3.7542	3.7543
34 40	3.7543	3.7544	3.7545	3.7546	3.7547	3.7547	3.7548	3.7549	3.7550	3.7550
34 50	3.7551	3.7552	3.7553	3.7553	3.7554	3.7555	3.7556	3.7556	3.7557	3.7558
1 35 0	3.7559	3.7560	3.7560	3.7561	3.7562	3.7563	3.7563	3.7564	3.7565	3.7566
35 10	3.7566	3.7567	3.7568	3.7569	3.7569	3.7570	3.7571	3.7572	3.7572	3.7573
35 20	3.7574	3.7575	3.7575	3.7576	3.7577	3.7578	3.7579	3.7579	3.7580	3.7581
35 30	3.7582	3.7582	3.7583	3.7584	3.7585	3.7585	3.7586	3.7587	3.7588	3.7588
35 40	3.7589	3.7590	3.7591	3.7591	3.7592	3.7593	3.7594	3.7594	3.7595	3.7596
35 50	3.7597	3.7597	3.7598	3.7599	3.7600	3.7600	3.7601	3.7602	3.7603	3.7603
1 36 0	3.7604	3.7605	3.7606	3.7606	3.7607	3.7608	3.7609	3.7609	3.7610	3.7611
36 10	3.7612	3.7613	3.7613	3.7614	3.7615	3.7616	3.7616	3.7617	3.7618	3.7619
36 20	3.7619	3.7620	3.7621	3.7622	3.7622	3.7623	3.7624	3.7625	3.7625	3.7626
36 30	3.7627	3.7628	3.7628	3.7629	3.7630	3.7631	3.7631	3.7632	3.7633	3.7634
36 40	3.7634	3.7635	3.7636	3.7637	3.7637	3.7638	3.7639	3.7640	3.7640	3.7641
36 50	3.7642	3.7643	3.7643	3.7644	3.7645	3.7645	3.7646	3.7647	3.7648	3.7648
1 37 0	3.7649	3.7650	3.7651	3.7651	3.7652	3.7653	3.7654	3.7654	3.7655	3.7656
37 10	3.7657	3.7657	3.7658	3.7659	3.7660	3.7660	3.7661	3.7662	3.7663	3.7663
37 20	3.7664	3.7665	3.7666	3.7666	3.7667	3.7668	3.7669	3.7669	3.7670	3.7671
37 30	3.7672	3.7672	3.7673	3.7674	3.7675	3.7675	3.7676	3.7677	3.7677	3.7678
37 40	3.7679	3.7680	3.7681	3.7681	3.7682	3.7683	3.7683	3.7684	3.7685	3.7686
37 50	3.7686	3.7687	3.7688	3.7689	3.7689	3.7690	3.7691	3.7692	3.7692	3.7693
1 38 0	3.7694	3.7695	3.7695	3.7696	3.7697	3.7697	3.7698	3.7699	3.7700	3.7700
38 10	3.7701	3.7702	3.7703	3.7703	3.7704	3.7705	3.7706	3.7706	3.7707	3.7708
38 20	3.7709	3.7709	3.7710	3.7711	3.7711	3.7712	3.7713	3.7714	3.7714	3.7715
38 30	3.7716	3.7717	3.7717	3.7718	3.7719	3.7720	3.7720	3.7721	3.7722	3.7722
38 40	3.7723	3.7724	3.7725	3.7725	3.7726	3.7727	3.7728	3.7728	3.7729	3.7730
38 50	3.7731	3.7731	3.7732	3.7733	3.7733	3.7734	3.7735	3.7736	3.7736	3.7737
1 39 0	3.7738	3.7739	3.7739	3.7740	3.7741	3.7742	3.7742	3.7743	3.7744	3.7744
39 10	3.7745	3.7746	3.7747	3.7747	3.7748	3.7749	3.7750	3.7750	3.7751	3.7752
39 20	3.7752	3.7753	3.7754	3.7755	3.7755	3.7756	3.7757	3.7758	3.7758	3.7759
39 30	3.7760	3.7760	3.7761	3.7762	3.7763	3.7763	3.7764	3.7765	3.7766	3.7766
39 40	3.7767	3.7768	3.7768	3.7769	3.7770	3.7771	3.7771	3.7772	3.7773	3.7774
39 50	3.7774	3.7775	3.7776	3.7776	3.7777	3.7778	3.7779	3.7779	3.7780	3.7781

7

TABLE IX.

LOGARITHMS OF SMALL ARCS IN SPACE OR TIME.

Arc.	0″	1″	2″	3″	4″	5″	6″	7″	8″	9″
1ʰ.40ᵐ. 0ˢ	3.7782	3.7782	3.7783	3.7784	3.7784	3.7785	3.7786	3.7787	3.7787	3.7788
40 10	3.7789	3.7789	3.7790	3.7791	3.7792	3.7792	3.7793	3.7794	3.7795	3.7795
40 20	3.7796	3.7797	3.7797	3.7798	3.7799	3.7800	3.7800	3.7801	3.7802	3.7802
40 30	3.7803	3.7804	3.7805	3.7805	3.7806	3.7807	3.7807	3.7808	3.7809	3.7810
40 40	3.7810	3.7811	3.7812	3.7813	3.7813	3.7814	3.7815	3.7815	3.7816	3.7817
40 50	3.7818	3.7818	3.7819	3.7820	3.7820	3.7821	3.7822	3.7823	3.7823	3.7824
1 41 0	3.7825	3.7825	3.7826	3.7827	3.7828	3.7828	3.7829	3.7830	3.7830	3.7831
41 10	3.7832	3.7833	3.7833	3.7834	3.7835	3.7835	3.7836	3.7837	3.7838	3.7838
41 20	3.7839	3.7840	3.7840	3.7841	3.7842	3.7843	3.7843	3.7844	3.7845	3.7845
41 30	3.7846	3.7847	3.7848	3.7848	3.7849	3.7850	3.7850	3.7851	3.7852	3.7853
41 40	3.7853	3.7854	3.7855	3.7855	3.7856	3.7857	3.7858	3.7858	3.7859	3.7860
41 50	3.7860	3.7861	3.7862	3.7863	3.7863	3.7864	3.7865	3.7865	3.7866	3.7867
1 42 0	3.7868	3.7868	3.7869	3.7870	3.7870	3.7871	3.7872	3.7872	3.7873	3.7874
42 10	3.7875	3.7875	3.7876	3.7877	3.7877	3.7878	3.7879	3.7880	3.7880	3.7881
42 20	3.7882	3.7882	3.7883	3.7884	3.7885	3.7885	3.7886	3.7887	3.7887	3.7888
42 30	3.7889	3.7889	3.7890	3.7891	3.7892	3.7892	3.7893	3.7894	3.7894	3.7895
42 40	3.7896	3.7897	3.7897	3.7898	3.7899	3.7899	3.7900	3.7901	3.7901	3.7902
42 50	3.7903	3.7904	3.7904	3.7905	3.7906	3.7906	3.7907	3.7908	3.7908	3.7909
1 43 0	3.7910	3.7911	3.7911	3.7912	3.7913	3.7913	3.7914	3.7915	3.7916	3.7916
43 10	3.7917	3.7918	3.7918	3.7919	3.7920	3.7920	3.7921	3.7922	3.7923	3.7923
43 20	3.7924	3.7925	3.7925	3.7926	3.7927	3.7927	3.7928	3.7929	3.7930	3.7930
43 30	3.7931	3.7932	3.7932	3.7933	3.7934	3.7934	3.7935	3.7936	3.7937	3.7937
43 40	3.7938	3.7939	3.7939	3.7940	3.7941	3.7941	3.7942	3.7943	3.7943	3.7944
43 50	3.7945	3.7946	3.7946	3.7947	3.7948	3.7948	3.7949	3.7950	3.7950	3.7951
1 44 0	3.7952	3.7953	3.7953	3.7954	3.7955	3.7955	3.7956	3.7957	3.7957	3.7958
44 10	3.7959	3.7959	3.7960	3.7961	3.7962	3.7962	3.7963	3.7964	3.7964	3.7965
44 20	3.7966	3.7966	3.7967	3.7968	3.7969	3.7969	3.7970	3.7971	3.7971	3.7972
44 30	3.7973	3.7973	3.7974	3.7975	3.7975	3.7976	3.7977	3.7978	3.7978	3.7979
44 40	3.7980	3.7980	3.7981	3.7982	3.7982	3.7983	3.7984	3.7984	3.7985	3.7986
44 50	3.7987	3.7987	3.7988	3.7989	3.7989	3.7990	3.7991	3.7991	3.7992	3.7993
1 45 0	3.7993	3.7994	3.7995	3.7995	3.7996	3.7997	3.7998	3.7998	3.7999	3.8000
45 10	3.8000	3.8001	3.8002	3.8002	3.8003	3.8004	3.8004	3.8005	3.8006	3.8006
45 20	3.8007	3.8008	3.8009	3.8009	3.8010	3.8011	3.8011	3.8012	3.8013	3.8013
45 30	3.8014	3.8015	3.8015	3.8016	3.8017	3.8017	3.8018	3.8019	3.8020	3.8020
45 40	3.8021	3.8022	3.8022	3.8023	3.8024	3.8024	3.8025	3.8026	3.8026	3.8027
45 50	3.8028	3.8028	3.8029	3.8030	3.8030	3.8031	3.8032	3.8033	3.8033	3.8034
1 46 0	3.8035	3.8035	3.8036	3.8036	3.8037	3.8038	3.8039	3.8039	3.8040	3.8041
46 10	3.8041	3.8042	3.8043	3.8043	3.8044	3.8045	3.8045	3.8046	3.8047	3.8048
46 20	3.8048	3.8049	3.8050	3.8050	3.8051	3.8052	3.8052	3.8053	3.8054	3.8054
46 30	3.8055	3.8056	3.8056	3.8057	3.8058	3.8058	3.8059	3.8060	3.8060	3.8061
46 40	3.8062	3.8062	3.8063	3.8064	3.8065	3.8065	3.8066	3.8067	3.8067	3.8068
46 50	3.8069	3.8069	3.8070	3.8071	3.8071	3.8072	3.8073	3.8073	3.8074	3.8075
1 47 0	3.8075	3.8076	3.8077	3.8077	3.8078	3.8079	3.8079	3.8080	3.8081	3.8081
47 10	3.8082	3.8083	3.8083	3.8084	3.8085	3.8085	3.8086	3.8087	3.8088	3.8088
47 20	3.8089	3.8090	3.8090	3.8091	3.8092	3.8092	3.8093	3.8094	3.8094	3.8095
47 30	3.8096	3.8096	3.8097	3.8098	3.8098	3.8099	3.8099	3.8100	3.8101	3.8102
47 40	3.8102	3.8103	3.8103	3.8104	3.8105	3.8106	3.8106	3.8107	3.8108	3.8108
47 50	3.8109	3.8110	3.8110	3.8111	3.8112	3.8112	3.8113	3.8114	3.8114	3.8115
1 48 0	3.8116	3.8116	3.8117	3.8118	3.8118	3.8119	3.8120	3.8120	3.8121	3.8122
48 10	3.8122	3.8123	3.8124	3.8124	3.8125	3.8126	3.8126	3.8127	3.8128	3.8128
48 20	3.8129	3.8130	3.8130	3.8131	3.8132	3.8132	3.8133	3.8134	3.8134	3.8135
48 30	3.8136	3.8136	3.8137	3.8138	3.8138	3.8139	3.8140	3.8140	3.8141	3.8142
48 40	3.8142	3.8143	3.8144	3.8144	3.8145	3.8146	3.8146	3.8147	3.8148	3.8148
48 50	3.8149	3.8150	3.8150	3.8151	3.8152	3.8152	3.8153	3.8154	3.8154	3.8155
1 49 0	3.8156	3.8156	3.8157	3.8158	3.8158	3.8159	3.8160	3.8160	3.8161	3.8162
49 10	3.8162	3.8163	3.8164	3.8164	3.8165	3.8166	3.8166	3.8167	3.8168	3.8168
49 20	3.8169	3.8170	3.8170	3.8171	3.8172	3.8172	3.8173	3.8174	3.8174	3.8175
49 30	3.8176	3.8176	3.8177	3.8178	3.8178	3.8179	3.8180	3.8180	3.8181	3.8182
49 40	3.8182	3.8183	3.8184	3.8184	3.8185	3.8185	3.8186	3.8187	3.8188	3.8188
49 50	3.8189	3.8190	3.8190	3.8191	3.8191	3.8192	3.8193	3.8193	3.8194	3.8195

TABLE IX.

LOGARITHMS OF SMALL ARCS IN SPACE OR TIME.

Arc.	0″	1″	2″	3″	4″	5″	6″	7″	8″	9″
1h.50m. 0s.	3.8195	3.8196	3.8197	3.8197	3.8198	3.8199	3.8199	3.8200	3.8201	3.8201
50 10	3.8202	3.8203	3.8203	3.8204	3.8205	3.8205	3.8206	3.8207	3.8207	3.8208
50 20	3.8209	3.8209	3.8210	3.8211	3.8211	3.8212	3.8213	3.8213	3.8214	3.8214
50 30	3.8215	3.8216	3.8216	3.8217	3.8218	3.8218	3.8219	3.8220	3.8220	3.8221
50 40	3.8222	3.8222	3.8223	3.8224	3.8224	3.8225	3.8226	3.8226	3.8227	3.8228
50 50	3.8228	3.8229	3.8230	3.8230	3.8231	3.8231	3.8232	3.8233	3.8233	3.8234
1 51 0	3.8235	3.8235	3.8236	3.8237	3.8237	3.8238	3.8239	3.8239	3.8240	3.8241
51 10	3.8241	3.8242	3.8243	3.8243	3.8244	3.8245	3.8245	3.8246	3.8246	3.8247
51 20	3.8248	3.8248	3.8249	3.8250	3.8250	3.8251	3.8252	3.8252	3.8253	3.8254
51 30	3.8254	3.8255	3.8256	3.8256	3.8257	3.8258	3.8258	3.8259	3.8259	3.8260
51 40	3.8261	3.8261	3.8262	3.8263	3.8263	3.8264	3.8265	3.8265	3.8266	3.8267
51 50	3.8267	3.8269	3.8269	3.8269	3.8270	3.8270	3.8271	3.8272	3.8272	3.8273
1 52 0	3.8274	3.8274	3.8275	3.8276	3.8276	3.8277	3.8278	3.8278	3.8279	3.8280
52 10	3.8280	3.8281	3.8281	3.8282	3.8283	3.8283	3.8284	3.8285	3.8285	3.8286
52 20	3.8287	3.8287	3.8288	3.8289	3.8289	3.8290	3.8290	3.8291	3.8292	3.8292
52 30	3.8293	3.8294	3.8294	3.8295	3.8296	3.8296	3.8297	3.8298	3.8298	3.8299
52 40	3.8299	3.8300	3.8301	3.8301	3.8302	3.8303	3.8303	3.8304	3.8305	3.8305
52 50	3.8306	3.8307	3.8307	3.8308	3.8308	3.8309	3.8310	3.8310	3.8311	3.8312
1 53 0	3.8312	3.8313	3.8314	3.8314	3.8315	3.8315	3.8316	3.8317	3.8317	3.8318
53 10	3.8319	3.8319	3.8320	3.8321	3.8321	3.8322	3.8323	3.8323	3.8324	3.8324
53 20	3.8325	3.8326	3.8326	3.8327	3.8328	3.8328	3.8329	3.8330	3.8330	3.8331
53 30	3.8331	3.8332	3.8333	3.8333	3.8334	3.8335	3.8335	3.8336	3.8337	3.8337
53 40	3.8338	3.8338	3.8339	3.8340	3.8340	3.8341	3.8342	3.8342	3.8343	3.8344
53 50	3.8344	3.8345	3.8345	3.8346	3.8347	3.8347	3.8348	3.8349	3.8349	3.8350
1 54 0	3.8351	3.8351	3.8352	3.8352	3.8353	3.8354	3.8354	3.8355	3.8356	3.8356
54 10	3.8357	3.8358	3.8358	3.8359	3.8359	3.8360	3.8361	3.8361	3.8362	3.8363
54 20	3.8363	3.8364	3.8365	3.8365	3.8366	3.8366	3.8367	3.8368	3.8368	3.8369
54 30	3.8370	3.8370	3.8371	3.8371	3.8372	3.8373	3.8373	3.8374	3.8375	3.8375
54 40	3.8376	3.8377	3.8377	3.8378	3.8378	3.8379	3.8380	3.8380	3.8381	3.8382
54 50	3.8382	3.8383	3.8383	3.8384	3.8385	3.8385	3.8386	3.8387	3.8387	3.8388
1 55 0	3.8388	3.8389	3.8390	3.8390	3.8391	3.8392	3.8392	3.8393	3.8394	3.8394
55 10	3.8395	3.8395	3.8396	3.8397	3.8397	3.8398	3.8399	3.8399	3.8400	3.8400
55 20	3.8401	3.8402	3.8402	3.8403	3.8404	3.8404	3.8405	3.8405	3.8406	3.8407
55 30	3.8407	3.8408	3.8409	3.8409	3.8410	3.8410	3.8411	3.8412	3.8412	3.8413
55 40	3.8414	3.8414	3.8415	3.8415	3.8416	3.8417	3.8417	3.8418	3.8419	3.8419
55 50	3.8420	3.8420	3.8421	3.8422	3.8422	3.8423	3.8424	3.8424	3.8425	3.8425
1 56 0	3.8426	3.8427	3.8427	3.8428	3.8429	3.8429	3.8430	3.8430	3.8431	3.8432
56 10	3.8432	3.8433	3.8434	3.8434	3.8435	3.8435	3.8436	3.8437	3.8437	3.8438
56 20	3.8439	3.8439	3.8440	3.8440	3.8441	3.8442	3.8442	3.8443	3.8444	3.8444
56 30	3.8445	3.8445	3.8446	3.8447	3.8447	3.8448	3.8448	3.8449	3.8450	3.8450
56 40	3.8451	3.8452	3.8452	3.8453	3.8453	3.8454	3.8455	3.8455	3.8456	3.8457
56 50	3.8457	3.8458	3.8458	3.8459	3.8460	3.8460	3.8461	3.8462	3.8462	3.8463
1 57 0	3.8463	3.8464	3.8465	3.8465	3.8466	3.8466	3.8467	3.8468	3.8468	3.8469
57 10	3.8470	3.8470	3.8471	3.8471	3.8472	3.8473	3.8473	3.8474	3.8474	3.8475
57 20	3.8476	3.8476	3.8477	3.8478	3.8478	3.8479	3.8479	3.8480	3.8481	3.8481
57 30	3.8482	3.8483	3.8483	3.8484	3.8484	3.8485	3.8486	3.8486	3.8487	3.8487
57 40	3.8488	3.8489	3.8489	3.8490	3.8491	3.8491	3.8492	3.8492	3.8493	3.8494
57 50	3.8494	3.8495	3.8495	3.8496	3.8497	3.8497	3.8498	3.8499	3.8499	3.8500
1 58 0	3.8500	3.8501	3.8502	3.8502	3.8503	3.8503	3.8504	3.8505	3.8505	3.8506
58 10	3.8506	3.8507	3.8508	3.8508	3.8509	3.8510	3.8510	3.8511	3.8511	3.8512
58 20	3.8513	3.8513	3.8514	3.8514	3.8515	3.8516	3.8516	3.8517	3.8517	3.8518
58 30	3.8519	3.8519	3.8520	3.8521	3.8521	3.8522	3.8522	3.8523	3.8524	3.8524
58 40	3.8525	3.8525	3.8526	3.8527	3.8527	3.8528	3.8528	3.8529	3.8530	3.8530
58 50	3.8531	3.8532	3.8532	3.8533	3.8533	3.8534	3.8535	3.8535	3.8536	3.8536
1 59 0	3.8537	3.8538	3.8538	3.8539	3.8539	3.8540	3.8541	3.8541	3.8542	3.8542
59 10	3.8543	3.8544	3.8544	3.8545	3.8545	3.8546	3.8547	3.8547	3.8548	3.8549
59 20	3.8549	3.8550	3.8550	3.8551	3.8552	3.8552	3.8553	3.8553	3.8554	3.8555
59 30	3.8555	3.8556	3.8556	3.8557	3.8558	3.8558	3.8559	3.8559	3.8560	3.8561
59 40	3.8561	3.8562	3.8562	3.8563	3.8564	3.8564	3.8565	3.8565	3.8566	3.8567
59 50	3.8567	3.8568	3.8568	3.8569	3.8570	3.8570	3.8571	3.8572	3.8572	3.8573

TABLE IX.

LOGARITHMS OF SMALL ARCS IN SPACE OR TIME.

Arc.		0″	1″	2″	3″	4″	5″	6″	7″	8″	9″
0	0	3.8573	3.8574	3.8575	3.8575	3.8576	3.8576	3.8577	3.8578	3.8578	3.8579
0	10	3.8579	3.8580	3.8581	3.8581	3.8582	3.8582	3.8583	3.8584	3.8584	3.8585
0	20	3.8585	3.8586	3.8587	3.8587	3.8588	3.8588	3.8589	3.8590	3.8590	3.8591
0	30	3.8591	3.8592	3.8593	3.8593	3.8594	3.8594	3.8595	3.8596	3.8596	3.8597
0	40	3.8597	3.8598	3.8599	3.8599	3.8600	3.8600	3.8601	3.8602	3.8602	3.8603
0	50	3.8603	3.8604	3.8605	3.8605	3.8606	3.8606	3.8607	3.8608	3.8608	3.8609
2 1	0	3.8609	3.8610	3.8611	3.8611	3.8612	3.8612	3.8613	3.8614	3.8614	3.8615
1	10	3.8615	3.8616	3.8617	3.8617	3.8618	3.8618	3.8619	3.8620	3.8620	3.8621
1	20	3.8621	3.8622	3.8623	3.8623	3.8624	3.8624	3.8625	3.8625	3.8626	3.8627
1	30	3.8627	3.8628	3.8628	3.8629	3.8630	3.8630	3.8631	3.8631	3.8632	3.8633
1	40	3.8633	3.8634	3.8634	3.8635	3.8636	3.8636	3.8637	3.8637	3.8638	3.8639
1	50	3.8639	3.8640	3.8640	3.8641	3.8642	3.8642	3.8643	3.8643	3.8644	3.8645
2 2	0	3.8645	3.8646	3.8646	3.8647	3.8647	3.8648	3.8649	3.8649	3.8650	3.8650
2	10	3.8651	3.8652	3.8652	3.8653	3.8653	3.8654	3.8655	3.8655	3.8656	3.8656
2	20	3.8657	3.8658	3.8658	3.8659	3.8659	3.8660	3.8661	3.8661	3.8662	3.8662
2	30	3.8663	3.8663	3.8664	3.8665	3.8665	3.8666	3.8666	3.8667	3.8668	3.8668
2	40	3.8669	3.8669	3.8670	3.8671	3.8671	3.8672	3.8672	3.8673	3.8673	3.8674
2	50	3.8675	3.8675	3.8676	3.8676	3.8677	3.8678	3.8678	3.8679	3.8679	3.8680
2 3	0	3.8681	3.8681	3.8682	3.8682	3.8683	3.8684	3.8684	3.8685	3.8685	3.8686
3	10	3.8686	3.8687	3.8688	3.8688	3.8689	3.8689	3.8690	3.8691	3.8691	3.8692
3	20	3.8692	3.8693	3.8693	3.8694	3.8695	3.8695	3.8696	3.8696	3.8697	3.8698
3	30	3.8698	3.8699	3.8699	3.8700	3.8701	3.8701	3.8702	3.8702	3.8703	3.8703
3	40	3.8704	3.8705	3.8705	3.8706	3.8706	3.8707	3.8708	3.8708	3.8709	3.8709
3	50	3.8710	3.8710	3.8711	3.8712	3.8712	3.8713	3.8713	3.8714	3.8715	3.8715
2 4	0	3.8716	3.8716	3.8717	3.8717	3.8718	3.8719	3.8719	3.8720	3.8720	3.8721
4	10	3.8722	3.8722	3.8723	3.8723	3.8724	3.8724	3.8725	3.8726	3.8726	3.8727
4	20	3.8727	3.8728	3.8729	3.8729	3.8730	3.8730	3.8731	3.8731	3.8732	3.8733
4	30	3.8733	3.8734	3.8734	3.8735	3.8736	3.8736	3.8737	3.8737	3.8738	3.8738
4	40	3.8739	3.8740	3.8740	3.8741	3.8741	3.8742	3.8742	3.8743	3.8744	3.8744
4	50	3.8745	3.8745	3.8746	3.8747	3.8747	3.8748	3.8748	3.8749	3.8749	3.8750
2 5	0	3.8751	3.8751	3.8752	3.8752	3.8753	3.8754	3.8754	3.8755	3.8755	3.8756
5	10	3.8756	3.8757	3.8758	3.8758	3.8759	3.8759	3.8760	3.8760	3.8761	3.8762
5	20	3.8762	3.8763	3.8763	3.8764	3.8764	3.8765	3.8766	3.8766	3.8767	3.8767
5	30	3.8768	3.8769	3.8769	3.8770	3.8770	3.8771	3.8771	3.8772	3.8773	3.8773
5	40	3.8774	3.8774	3.8775	3.8775	3.8776	3.8777	3.8777	3.8778	3.8778	3.8779
5	50	3.8779	3.8780	3.8781	3.8781	3.8782	3.8782	3.8783	3.8784	3.8784	3.8785
2 6	0	3.8785	3.8786	3.8786	3.8787	3.8788	3.8788	3.8789	3.8789	3.8790	3.8790
6	10	3.8791	3.8792	3.8792	3.8793	3.8793	3.8794	3.8794	3.8795	3.8796	3.8796
6	20	3.8797	3.8797	3.8798	3.8798	3.8799	3.8800	3.8800	3.8801	3.8801	3.8802
6	30	3.8802	3.8803	3.8804	3.8804	3.8805	3.8805	3.8806	3.8806	3.8807	3.8808
6	40	3.8808	3.8809	3.8809	3.8810	3.8810	3.8811	3.8812	3.8812	3.8813	3.8813
6	50	3.8814	3.8814	3.8815	3.8816	3.8816	3.8817	3.8817	3.8818	3.8818	3.8819
2 7	0	3.8820	3.8820	3.8821	3.8821	3.8822	3.8822	3.8823	3.8824	3.8824	3.8825
7	10	3.8825	3.8826	3.8826	3.8827	3.8828	3.8828	3.8829	3.8829	3.8830	3.8830
7	20	3.8831	3.8832	3.8832	3.8833	3.8833	3.8834	3.8834	3.8835	3.8835	3.8836
7	30	3.8837	3.8837	3.8838	3.8838	3.8839	3.8839	3.8840	3.8841	3.8841	3.8842
7	40	3.8842	3.8843	3.8843	3.8844	3.8845	3.8845	3.8846	3.8846	3.8847	3.8847
7	50	3.8848	3.8849	3.8849	3.8850	3.8850	3.8851	3.8851	3.8852	3.8852	3.8853
2 8	0	3.8854	3.8854	3.8855	3.8855	3.8856	3.8856	3.8857	3.8858	3.8858	3.8859
8	10	3.8859	3.8860	3.8860	3.8861	3.8862	3.8862	3.8863	3.8863	3.8864	3.8864
8	20	3.8865	3.8865	3.8866	3.8867	3.8867	3.8868	3.8868	3.8869	3.8869	3.8870
8	30	3.8871	3.8871	3.8872	3.8872	3.8873	3.8873	3.8874	3.8874	3.8875	3.8876
8	40	3.8876	3.8877	3.8877	3.8878	3.8878	3.8879	3.8880	3.8880	3.8881	3.8881
8	50	3.8882	3.8882	3.8883	3.8883	3.8884	3.8885	3.8885	3.8886	3.8886	3.8887
2 9	0	3.8897	3.8888	3.8889	3.8889	3.8890	3.8890	3.8891	3.8891	3.8892	3.8892
9	10	3.8893	3.8894	3.8894	3.8895	3.8895	3.8896	3.8896	3.8897	3.8897	3.8898
9	20	3.8899	3.8899	3.8900	3.8900	3.8901	3.8901	3.8902	3.8903	3.8903	3.8904
9	30	3.8904	3.8905	3.8905	3.8906	3.8906	3.8907	3.8908	3.8908	3.8909	3.8909
9	40	3.8910	3.8910	3.8911	3.8911	3.8912	3.8912	3.8913	3.8914	3.8914	3.8915
9	50	3.8915	3.8916	3.8916	3.8917	3.8918	3.8918	3.8919	3.8919	3.8920	3.8920

TABLE IX.

LOGARITHMS OF SMALL ARCS IN SPACE OR TIME.

Arc.	0″	1″	2″	3″	4″	5″	6″	7″	8″	9″
2ʰ 10ᵐ 0ˢ	3.8921	3.8922	3.8922	3.8923	3.8923	3.8924	3.8924	3.8925	3.8925	3.8926
10 10	3.8927	3.8927	3.8928	3.8928	3.8929	3.8929	3.8930	3.8930	3.8931	3.8932
10 20	3.8932	3.8933	3.8933	3.8934	3.8934	3.8935	3.8935	3.8936	3.8937	3.8937
10 30	3.8938	3.8938	3.8939	3.8939	3.8940	3.8940	3.8941	3.8941	3.8942	3.8943
10 40	3.8943	3.8944	3.8944	3.8945	3.8945	3.8946	3.8946	3.8947	3.8948	3.8948
10 50	3.8949	3.8949	3.8950	3.8950	3.8951	3.8951	3.8952	3.8953	3.8953	3.8954
2 11 0	3.8954	3.8955	3.8955	3.8956	3.8956	3.8957	3.8958	3.8958	3.8959	3.8959
11 10	3.8960	3.8960	3.8961	3.8961	3.8962	3.8963	3.8963	3.8964	3.8964	3.8965
11 20	3.8965	3.8966	3.8966	3.8967	3.8967	3.8968	3.8969	3.8969	3.8970	3.8970
11 30	3.8971	3.8971	3.8972	3.8972	3.8973	3.8974	3.8974	3.8975	3.8975	3.8976
11 40	3.8976	3.8977	3.8977	3.8978	3.8978	3.8979	3.8980	3.8980	3.8981	3.8981
11 50	3.8982	3.8982	3.8982	3.8983	3.8984	3.8985	3.8985	3.8986	3.8986	3.8987
2 12 0	3.8987	3.8988	3.8988	3.8989	3.8989	3.8990	3.8991	3.8991	3.8992	3.8992
12 10	3.8993	3.8993	3.8994	3.8994	3.8995	3.8995	3.8996	3.8997	3.8997	3.8998
12 20	3.8998	3.8999	3.8999	3.9000	3.9000	3.9001	3.9001	3.9002	3.9003	3.9003
12 30	3.9004	3.9004	3.9005	3.9005	3.9006	3.9006	3.9007	3.9007	3.9008	3.9009
12 40	3.9009	3.9010	3.9010	3.9011	3.9011	3.9012	3.9012	3.9013	3.9013	3.9014
12 50	3.9015	3.9015	3.9016	3.9016	3.9017	3.9017	3.9018	3.9018	3.9019	3.9019
2 13 0	3.9020	3.9021	3.9021	3.9022	3.9022	3.9023	3.9023	3.9024	3.9024	3.9025
13 10	3.9025	3.9026	3.9027	3.9027	3.9028	3.9028	3.9029	3.9030	3.9030	3.9031
13 20	3.9031	3.9031	3.9032	3.9033	3.9033	3.9034	3.9034	3.9035	3.9035	3.9036
13 30	3.9036	3.9037	3.9037	3.9038	3.9038	3.9039	3.9040	3.9040	3.9041	3.9041
13 40	3.9042	3.9042	3.9043	3.9043	3.9044	3.9044	3.9045	3.9046	3.9046	3.9047
13 50	3.9047	3.9048	3.9048	3.9049	3.9049	3.9050	3.9050	3.9051	3.9051	3.9052
2 14 0	3.9053	3.9053	3.9054	3.9054	3.9055	3.9055	3.9056	3.9056	3.9057	3.9057
14 10	3.9058	3.9058	3.9059	3.9060	3.9060	3.9061	3.9061	3.9062	3.9062	3.9063
14 20	3.9063	3.9064	3.9064	3.9065	3.9065	3.9066	3.9067	3.9067	3.9068	3.9068
14 30	3.9069	3.9069	3.9070	3.9070	3.9071	3.9071	3.9072	3.9073	3.9073	3.9074
14 40	3.9074	3.9075	3.9075	3.9076	3.9076	3.9077	3.9077	3.9078	3.9078	3.9079
14 50	3.9079	3.9080	3.9081	3.9081	3.9082	3.9082	3.9083	3.9083	3.9084	3.9084
2 15 0	3.9085	3.9085	3.9086	3.9086	3.9087	3.9088	3.9088	3.9089	3.9089	3.9090
15 10	3.9090	3.9091	3.9091	3.9092	3.9092	3.9093	3.9093	3.9094	3.9094	3.9095
15 20	3.9096	3.9096	3.9097	3.9097	3.9098	3.9098	3.9099	3.9099	3.9100	3.9100
15 30	3.9101	3.9101	3.9102	3.9102	3.9103	3.9104	3.9104	3.9105	3.9105	3.9106
15 40	3.9106	3.9107	3.9107	3.9108	3.9108	3.9109	3.9109	3.9110	3.9111	3.9111
15 50	3.9112	3.9112	3.9113	3.9113	3.9114	3.9114	3.9115	3.9115	3.9116	3.9116
2 16 0	3.9117	3.9117	3.9118	3.9118	3.9119	3.9120	3.9120	3.9121	3.9121	3.9122
16 10	3.9122	3.9123	3.9123	3.9124	3.9124	3.9125	3.9125	3.9126	3.9126	3.9127
16 20	3.9128	3.9128	3.9129	3.9129	3.9130	3.9130	3.9131	3.9131	3.9132	3.9132
16 30	3.9133	3.9133	3.9134	3.9134	3.9135	3.9135	3.9136	3.9137	3.9137	3.9138
16 40	3.9138	3.9139	3.9139	3.9140	3.9140	3.9141	3.9141	3.9142	3.9142	3.9143
16 50	3.9143	3.9144	3.9144	3.9145	3.9145	3.9146	3.9146	3.9147	3.9147	3.9148
2 17 0	3.9149	3.9149	3.9150	3.9150	3.9151	3.9151	3.9152	3.9152	3.9153	3.9153
17 10	3.9154	3.9155	3.9155	3.9156	3.9156	3.9157	3.9157	3.9158	3.9158	3.9159
17 20	3.9159	3.9160	3.9160	3.9161	3.9161	3.9162	3.9162	3.9163	3.9163	3.9164
17 30	3.9165	3.9165	3.9166	3.9166	3.9167	3.9167	3.9168	3.9168	3.9169	3.9169
17 40	3.9170	3.9170	3.9171	3.9171	3.9172	3.9172	3.9173	3.9173	3.9174	3.9175
17 50	3.9175	3.9176	3.9176	3.9177	3.9177	3.9178	3.9178	3.9179	3.9179	3.9180
2 18 0	3.9180	3.9181	3.9181	3.9182	3.9182	3.9183	3.9183	3.9184	3.9184	3.9185
18 10	3.9186	3.9186	3.9187	3.9187	3.9198	3.9188	3.9189	3.9189	3.9190	3.9190
18 20	3.9191	3.9191	3.9192	3.9192	3.9193	3.9193	3.9194	3.9194	3.9195	3.9195
18 30	3.9196	3.9197	3.9197	3.9198	3.9198	3.9199	3.9199	3.9200	3.9200	3.9201
18 40	3.9201	3.9202	3.9202	3.9203	3.9203	3.9204	3.9204	3.9205	3.9205	3.9206
18 50	3.9206	3.9207	3.9207	3.9208	3.9209	3.9209	3.9210	3.9210	3.9211	3.9211
2 19 0	3.9212	3.9212	3.9213	3.9213	3.9214	3.9214	3.9215	3.9215	3.9216	3.9216
19 10	3.9217	3.9217	3.9218	3.9218	3.9219	3.9220	3.9220	3.9221	3.9221	3.9222
19 20	3.9222	3.9223	3.9223	3.9224	3.9224	3.9225	3.9225	3.9226	3.9226	3.9227
19 30	3.9227	3.9228	3.9228	3.9229	3.9229	3.9230	3.9230	3.9231	3.9231	3.9232
19 40	3.9232	3.9233	3.9233	3.9234	3.9235	3.9235	3.9236	3.9236	3.9237	3.9237
19 50	3.9238	3.9238	3.9239	3.9239	3.9240	3.9240	3.9241	3.9241	3.9242	3.9242

TABLE IX.

LOGARITHMS OF SMALL ARCS IN SPACE OR TIME.

Arc.	0	1	2	3	4	5	6	7	8	9
2h 20m 0s	3.9243	3.9243	3.9244	3.9244	3.9245	3.9245	3.9246	3.9246	3.9247	3.9247
20 10	3.9248	3.9248	3.9249	3.9250	3.9250	3.9251	3.9251	3.9252	3.9252	3.9253
20 20	3.9253	3.9254	3.9254	3.9255	3.9255	3.9256	3.9256	3.9257	3.9257	3.9258
20 30	3.9258	3.9259	3.9259	3.9260	3.9260	3.9261	3.9261	3.9262	3.9262	3.9263
20 40	3.9263	3.9264	3.9264	3.9265	3.9265	3.9266	3.9267	3.9267	3.9268	3.9268
20 50	3.9269	3.9269	3.9270	3.9270	3.9271	3.9271	3.9272	3.9272	3.9273	3.9273
2 21 0	3.9274	3.9274	3.9275	3.9275	3.9276	3.9276	3.9277	3.9277	3.9278	3.9278
21 10	3.9279	3.9279	3.9280	3.9280	3.9281	3.9281	3.9282	3.9282	3.9283	3.9283
21 20	3.9284	3.9284	3.9285	3.9285	3.9286	3.9287	3.9287	3.9288	3.9288	3.9289
21 30	3.9289	3.9290	3.9290	3.9291	3.9291	3.9292	3.9292	3.9293	3.9293	3.9294
21 40	3.9294	3.9295	3.9295	3.9296	3.9296	3.9297	3.9297	3.9298	3.9298	3.9299
21 50	3.9299	3.9300	3.9300	3.9301	3.9301	3.9302	3.9302	3.9303	3.9303	3.9304
2 22 0	3.9304	3.9305	3.9305	3.9306	3.9306	3.9307	3.9307	3.9308	3.9308	3.9309
22 10	3.9309	3.9310	3.9311	3.9311	3.9312	3.9312	3.9313	3.9313	3.9314	3.9314
22 20	3.9315	3.9315	3.9316	3.9316	3.9317	3.9317	3.9318	3.9318	3.9319	3.9319
22 30	3.9320	3.9320	3.9321	3.9321	3.9322	3.9322	3.9323	3.9323	3.9324	3.9324
22 40	3.9325	3.9325	3.9326	3.9326	3.9327	3.9327	3.9328	3.9328	3.9329	3.9329
22 50	3.9330	3.9330	3.9331	3.9331	3.9332	3.9332	3.9333	3.9333	3.9334	3.9334
2 23 0	3.9335	3.9335	3.9336	3.9336	3.9337	3.9337	3.9338	3.9338	3.9339	3.9339
23 10	3.9340	3.9340	3.9341	3.9341	3.9342	3.9342	3.9343	3.9343	3.9344	3.9344
23 20	3.9345	3.9345	3.9346	3.9346	3.9347	3.9348	3.9348	3.9349	3.9349	3.9350
23 30	3.9350	3.9351	3.9351	3.9352	3.9352	3.9353	3.9353	3.9354	3.9354	3.9355
23 40	3.9355	3.9356	3.9356	3.9357	3.9357	3.9358	3.9358	3.9359	3.9359	3.9360
23 50	3.9360	3.9361	3.9361	3.9362	3.9362	3.9363	3.9363	3.9364	3.9364	3.9365
2 24 0	3.9365	3.9366	3.9366	3.9367	3.9367	3.9368	3.9368	3.9369	3.9369	3.9370
24 10	3.9370	3.9371	3.9371	3.9372	3.9372	3.9373	3.9373	3.9374	3.9374	3.9375
24 20	3.9375	3.9376	3.9376	3.9377	3.9377	3.9378	3.9378	3.9379	3.9379	3.9380
24 30	3.9380	3.9381	3.9381	3.9382	3.9382	3.9383	3.9383	3.9384	3.9384	3.9385
24 40	3.9385	3.9386	3.9386	3.9387	3.9387	3.9388	3.9388	3.9389	3.9389	3.9390
24 50	3.9390	3.9391	3.9391	3.9392	3.9392	3.9393	3.9393	3.9394	3.9394	3.9395
2 25 0	3.9395	3.9396	3.9396	3.9397	3.9397	3.9398	3.9398	3.9399	3.9399	3.9400
25 10	3.9400	3.9401	3.9401	3.9402	3.9402	3.9403	3.9403	3.9404	3.9404	3.9405
25 20	3.9405	3.9406	3.9406	3.9407	3.9407	3.9408	3.9408	3.9409	3.9409	3.9410
25 30	3.9410	3.9411	3.9411	3.9412	3.9412	3.9413	3.9413	3.9414	3.9414	3.9415
25 40	3.9415	3.9416	3.9416	3.9417	3.9417	3.9418	3.9418	3.9419	3.9419	3.9420
25 50	3.9420	3.9421	3.9421	3.9422	3.9422	3.9423	3.9423	3.9424	3.9424	3.9425
2 26 0	3.9425	3.9426	3.9426	3.9427	3.9427	3.9428	3.9428	3.9429	3.9429	3.9430
26 10	3.9430	3.9430	3.9431	3.9431	3.9432	3.9432	3.9433	3.9433	3.9434	3.9434
26 20	3.9435	3.9435	3.9436	3.9436	3.9437	3.9437	3.9438	3.9438	3.9439	3.9439
26 30	3.9440	3.9440	3.9441	3.9441	3.9442	3.9442	3.9443	3.9443	3.9444	3.9444
26 40	3.9445	3.9445	3.9446	3.9446	3.9447	3.9447	3.9448	3.9448	3.9449	3.9449
26 50	3.9450	3.9450	3.9451	3.9451	3.9452	3.9452	3.9453	3.9453	3.9454	3.9454
2 27 0	3.9455	3.9455	3.9456	3.9456	3.9457	3.9457	3.9458	3.9458	3.9459	3.9459
27 10	3.9460	3.9460	3.9461	3.9461	3.9462	3.9462	3.9463	3.9463	3.9464	3.9464
27 20	3.9465	3.9465	3.9466	3.9466	3.9466	3.9467	3.9467	3.9468	3.9468	3.9469
27 30	3.9469	3.9470	3.9470	3.9471	3.9471	3.9472	3.9472	3.9473	3.9473	3.9474
27 40	3.9474	3.9475	3.9475	3.9476	3.9476	3.9477	3.9477	3.9478	3.9478	3.9479
27 50	3.9479	3.9480	3.9480	3.9481	3.9481	3.9482	3.9482	3.9483	3.9483	3.9484
2 28 0	3.9484	3.9485	3.9485	3.9486	3.9486	3.9487	3.9487	3.9488	3.9488	3.9489
28 10	3.9489	3.9490	3.9490	3.9490	3.9491	3.9491	3.9492	3.9492	3.9493	3.9493
28 20	3.9494	3.9494	3.9495	3.9495	3.9496	3.9496	3.9497	3.9497	3.9498	3.9498
28 30	3.9499	3.9499	3.9500	3.9500	3.9501	3.9501	3.9502	3.9502	3.9503	3.9503
28 40	3.9504	3.9504	3.9505	3.9505	3.9506	3.9506	3.9507	3.9507	3.9508	3.9508
28 50	3.9509	3.9509	3.9509	3.9510	3.9510	3.9511	3.9511	3.9512	3.9512	3.9513
2 29 0	3.9513	3.9514	3.9514	3.9515	3.9515	3.9516	3.9516	3.9517	3.9517	3.9518
29 10	3.9518	3.9519	3.9519	3.9520	3.9520	3.9521	3.9521	3.9522	3.9522	3.9523
29 20	3.9523	3.9524	3.9524	3.9525	3.9525	3.9526	3.9526	3.9527	3.9527	3.9528
29 30	3.9528	3.9528	3.9529	3.9529	3.9530	3.9530	3.9531	3.9531	3.9532	3.9532
29 40	3.9533	3.9533	3.9534	3.9534	3.9535	3.9535	3.9536	3.9536	3.9537	3.9537
29 50	3.9538	3.9538	3.9539	3.9539	3.9540	3.9540	3.9540	3.9541	3.9541	3.9542

TABLE IX.

LOGARITHMS OF SMALL ARCS IN SPACE OR TIME.

Arc.	0″	1″	2″	3″	4″	5″	6″	7″	8″	9″
2ʰ 30ᵐ 0ˢ	3.9542	3.9543	3.9543	3.9544	3.9544	3.9545	3.9545	3.9546	3.9546	3.9547
30 10	3.9547	3.9548	3.9548	3.9549	3.9549	3.9550	3.9550	3.9551	3.9551	3.9552
30 20	3.9552	3.9553	3.9553	3.9554	3.9554	3.9554	3.9555	3.9555	3.9556	3.9556
30 30	3.9557	3.9557	3.9558	3.9558	3.9559	3.9559	3.9560	3.9560	3.9561	3.9561
30 40	3.9562	3.9562	3.9563	3.9563	3.9564	3.9564	3.9565	3.9565	3.9566	3.9566
30 50	3.9566	3.9567	3.9567	3.9568	3.9568	3.9569	3.9569	3.9570	3.9570	3.9571
2 31 0	3.9571	3.9572	3.9572	3.9573	3.9573	3.9574	3.9574	3.9575	3.9575	3.9576
31 10	3.9576	3.9577	3.9577	3.9578	3.9578	3.9578	3.9579	3.9579	3.9580	3.9580
31 20	3.9581	3.9581	3.9582	3.9582	3.9583	3.9583	3.9584	3.9584	3.9585	3.9585
31 30	3.9586	3.9586	3.9587	3.9587	3.9588	3.9588	3.9589	3.9589	3.9589	3.9590
31 40	3.9590	3.9591	3.9591	3.9592	3.9592	3.9593	3.9593	3.9594	3.9594	3.9595
31 50	3.9595	3.9596	3.9596	3.9597	3.9597	3.9598	3.9598	3.9599	3.9599	3.9599
2 32 0	3.9600	3.9600	3.9601	3.9601	3.9602	3.9602	3.9603	3.9603	3.9604	3.9604
32 10	3.9605	3.9605	3.9606	3.9606	3.9607	3.9607	3.9608	3.9608	3.9609	3.9609
32 20	3.9609	3.9610	3.9610	3.9611	3.9611	3.9612	3.9612	3.9613	3.9613	3.9614
32 30	3.9614	3.9615	3.9615	3.9616	3.9616	3.9617	3.9617	3.9618	3.9618	3.9618
32 40	3.9619	3.9619	3.9620	3.9620	3.9621	3.9621	3.9622	3.9622	3.9623	3.9623
32 50	3.9624	3.9624	3.9625	3.9625	3.9626	3.9626	3.9627	3.9627	3.9627	3.9628
2 33 0	3.9628	3.9629	3.9629	3.9630	3.9630	3.9631	3.9631	3.9632	3.9632	3.9633
33 10	3.9633	3.9634	3.9634	3.9634	3.9635	3.9635	3.9636	3.9636	3.9637	3.9637
33 20	3.9638	3.9638	3.9639	3.9639	3.9640	3.9640	3.9641	3.9641	3.9642	3.9642
33 30	3.9642	3.9643	3.9643	3.9644	3.9644	3.9645	3.9645	3.9646	3.9646	3.9647
33 40	3.9647	3.9648	3.9648	3.9649	3.9649	3.9650	3.9650	3.9651	3.9651	3.9652
33 50	3.9652	3.9653	3.9653	3.9653	3.9654	3.9654	3.9655	3.9655	3.9656	3.9656
2 34 0	3.9657	3.9657	3.9658	3.9658	3.9658	3.9659	3.9659	3.9660	3.9660	3.9661
34 10	3.9661	3.9662	3.9662	3.9663	3.9663	3.9664	3.9664	3.9665	3.9665	3.9665
34 20	3.9666	3.9666	3.9667	3.9667	3.9668	3.9668	3.9669	3.9669	3.9670	3.9670
34 30	3.9671	3.9671	3.9672	3.9672	3.9672	3.9673	3.9673	3.9674	3.9674	3.9675
34 40	3.9675	3.9676	3.9676	3.9677	3.9677	3.9678	3.9678	3.9679	3.9679	3.9680
34 50	3.9680	3.9681	3.9681	3.9682	3.9682	3.9682	3.9683	3.9683	3.9684	3.9684
2 35 0	3.9685	3.9685	3.9686	3.9686	3.9687	3.9687	3.9688	3.9688	3.9689	3.9689
35 10	3.9689	3.9690	3.9690	3.9691	3.9691	3.9692	3.9692	3.9693	3.9693	3.9694
35 20	3.9694	3.9695	3.9695	3.9696	3.9696	3.9696	3.9697	3.9697	3.9698	3.9698
35 30	3.9699	3.9699	3.9700	3.9700	3.9701	3.9701	3.9702	3.9702	3.9703	3.9703
35 40	3.9703	3.9704	3.9704	3.9705	3.9705	3.9706	3.9706	3.9707	3.9707	3.9708
35 50	3.9708	3.9709	3.9709	3.9710	3.9710	3.9710	3.9711	3.9711	3.9712	3.9712
2 36 0	3.9713	3.9713	3.9714	3.9714	3.9714	3.9715	3.9715	3.9716	3.9716	3.9717
36 10	3.9717	3.9718	3.9718	3.9719	3.9719	3.9720	3.9720	3.9721	3.9721	3.9722
36 20	3.9722	3.9722	3.9723	3.9723	3.9724	3.9724	3.9725	3.9725	3.9726	3.9726
36 30	3.9727	3.9727	3.9728	3.9728	3.9729	3.9729	3.9729	3.9730	3.9730	3.9731
36 40	3.9731	3.9732	3.9732	3.9733	3.9733	3.9734	3.9734	3.9735	3.9735	3.9735
36 50	3.9736	3.9736	3.9737	3.9737	3.9738	3.9738	3.9739	3.9739	3.9740	3.9740
2 37 0	3.9741	3.9741	3.9741	3.9742	3.9742	3.9743	3.9743	3.9744	3.9744	3.9745
37 10	3.9745	3.9746	3.9746	3.9746	3.9747	3.9747	3.9748	3.9748	3.9749	3.9749
37 20	3.9750	3.9750	3.9751	3.9751	3.9752	3.9752	3.9752	3.9753	3.9753	3.9754
37 30	3.9754	3.9755	3.9755	3.9756	3.9756	3.9757	3.9757	3.9758	3.9758	3.9758
37 40	3.9759	3.9759	3.9760	3.9760	3.9761	3.9761	3.9762	3.9762	3.9763	3.9763
37 50	3.9763	3.9764	3.9764	3.9765	3.9765	3.9766	3.9766	3.9767	3.9767	3.9768
2 38 0	3.9768	3.9769	3.9769	3.9769	3.9770	3.9770	3.9771	3.9771	3.9772	3.9772
38 10	3.9773	3.9773	3.9774	3.9774	3.9774	3.9775	3.9775	3.9776	3.9776	3.9777
38 20	3.9777	3.9778	3.9778	3.9779	3.9779	3.9779	3.9780	3.9780	3.9781	3.9781
38 30	3.9782	3.9782	3.9783	3.9783	3.9784	3.9784	3.9785	3.9785	3.9785	3.9786
38 40	3.9786	3.9787	3.9787	3.9786	3.9788	3.9789	3.9889	3.9790	3.9790	3.9790
38 50	3.9791	3.9791	3.9792	3.9792	3.9793	3.9793	3.9794	3.9794	3.9795	3.9795
2 39 0	3.9795	3.9796	3.9796	3.9797	3.9797	3.9798	3.9798	3.9799	3.9799	3.9800
39 10	3.9800	3.9800	3.9801	3.9801	3.9802	3.9802	3.9803	3.9803	3.9804	3.9804
39 20	3.9805	3.9805	3.9805	3.9806	3.9806	3.9807	3.9807	3.9808	3.9808	3.9809
39 30	3.9809	3.9810	3.9810	3.9810	3.9811	3.9811	3.9812	3.9812	3.9813	3.9813
39 40	3.9814	3.9814	3.9815	3.9815	3.9815	3.9816	3.9816	3.9817	3.9817	3.9818
39 50	3.9818	3.9819	3.9819	3.9819	3.9820	3.9820	3.9821	3.9821	3.9822	3.9822

TABLE IX.

LOGARITHMS OF SMALL ARCS IN SPACE OR TIME.

Arc.	0″	1″	2″	3″	4″	5″	6″	7″	8″	9″
2ʰ.40ᵐ. 0ˢ.	3.9823	3.9823	3.9824	3.9824	3.9825	3.9825	3.9825	3.9826	3.9826	3.9827
40 10	3.9827	3.9828	3.9828	3.9829	3.9829	3.9829	3.9830	3.9830	3.9831	3.9831
40 20	3.9832	3.9832	3.9833	3.9833	3.9834	3.9834	3.9834	3.9835	3.9835	3.9836
40 30	3.9836	3.9837	3.9837	3.9837	3.9838	3.9838	3.9839	3.9839	3.9840	3.9840
40 40	3.9841	3.9841	3.9842	3.9842	3.9843	3.9843	3.9843	3.9844	3.9844	3.9845
40 50	3.9845	3.9846	3.9846	3.9847	3.9847	3.9848	3.9848	3.9848	3.9849	3.9849
2 41 0	3.9850	3.9850	3.9851	3.9851	3.9852	3.9852	3.9852	3.9853	3.9853	3.9854
41 10	3.9854	3.9855	3.9855	3.9856	3.9856	3.9857	3.9857	3.9857	3.9858	3.9858
41 20	3.9859	3.9859	3.9860	3.9860	3.9861	3.9861	3.9861	3.9862	3.9862	3.9863
41 30	3.9863	3.9864	3.9864	3.9865	3.9865	3.9865	3.9866	3.9866	3.9867	3.9867
41 40	3.9868	3.9868	3.9869	3.9869	3.9870	3.9870	3.9870	3.9871	3.9871	3.9872
41 50	3.9872	3.9873	3.9873	3.9874	3.9874	3.9874	3.9875	3.9875	3.9876	3.9876
2 42 0	3.9877	3.9877	3.9878	3.9878	3.9878	3.9879	3.9879	3.9880	3.9880	3.9881
42 10	3.9881	3.9882	3.9882	3.9882	3.9883	3.9883	3.9884	3.9884	3.9885	3.9885
42 20	3.9886	3.9886	3.9886	3.9887	3.9887	3.9888	3.9888	3.9889	3.9889	3.9890
42 30	3.9890	3.9890	3.9891	3.9891	3.9892	3.9892	3.9893	3.9893	3.9894	3.9894
42 40	3.9894	3.9895	3.9895	3.9896	3.9896	3.9897	3.9897	3.9898	3.9898	3.9898
42 50	3.9899	3.9899	3.9900	3.9900	3.9901	3.9901	3.9902	3.9902	3.9903	3.9903
2 43 0	3.9903	3.9904	3.9904	3.9905	3.9905	3.9906	3.9906	3.9906	3.9907	3.9907
43 10	3.9908	3.9908	3.9909	3.9909	3.9910	3.9910	3.9910	3.9911	3.9911	3.9912
43 20	3.9912	3.9913	3.9913	3.9914	3.9914	3.9914	3.9915	3.9915	3.9916	3.9916
43 30	3.9917	3.9917	3.9918	3.9918	3.9918	3.9919	3.9919	3.9920	3.9920	3.9921
43 40	3.9921	3.9922	3.9922	3.9922	3.9923	3.9923	3.9924	3.9924	3.9925	3.9925
43 50	3.9926	3.9926	3.9926	3.9927	3.9927	3.9928	3.9928	3.9929	3.9929	3.9930
2 44 0	3.9930	3.9930	3.9931	3.9931	3.9932	3.9932	3.9933	3.9933	3.9933	3.9934
44 10	3.9934	3.9935	3.9935	3.9936	3.9936	3.9937	3.9937	3.9937	3.9938	3.9938
44 20	3.9939	3.9939	3.9940	3.9940	3.9941	3.9941	3.9941	3.9942	3.9942	3.9943
44 30	3.9943	3.9944	3.9944	3.9944	3.9945	3.9945	3.9946	3.9946	3.9947	3.9947
44 40	3.9948	3.9948	3.9948	3.9949	3.9949	3.9950	3.9950	3.9951	3.9951	3.9952
44 50	3.9952	3.9952	3.9953	3.9953	3.9954	3.9954	3.9955	3.9955	3.9955	3.9956
2 45 0	3.9956	3.9957	3.9957	3.9958	3.9958	3.9959	3.9959	3.9959	3.9960	3.9960
45 10	3.9961	3.9961	3.9962	3.9962	3.9962	3.9963	3.9963	3.9964	3.9964	3.9965
45 20	3.9965	3.9966	3.9966	3.9966	3.9967	3.9967	3.9968	3.9968	3.9969	3.9969
45 30	3.9969	3.9970	3.9970	3.9971	3.9971	3.9972	3.9972	3.9973	3.9973	3.9973
45 40	3.9974	3.9974	3.9975	3.9975	3.9976	3.9976	3.9976	3.9977	3.9977	3.9978
45 50	3.9978	3.9979	3.9979	3.9980	3.9980	3.9980	3.9981	3.9981	3.9982	3.9982
2 46 0	3.9983	3.9983	3.9983	3.9984	3.9984	3.9985	3.9985	3.9986	3.9986	3.9987
46 10	3.9987	3.9987	3.9988	3.9988	3.9989	3.9989	3.9990	3.9990	3.9990	3.9991
46 20	3.9991	3.9992	3.9992	3.9993	3.9993	3.9993	3.9994	3.9994	3.9995	3.9995
46 30	3.9996	3.9996	3.9997	3.9997	3.9997	3.9998	3.9998	3.9999	3.9999	4.0000
46 40	4.0000	4.0000	4.0001	4.0001	4.0002	4.0002	4.0003	4.0003	4.0003	4.0004
46 50	4.0004	4.0005	4.0005	4.0006	4.0006	4.0007	4.0007	4.0007	4.0008	4.0008
2 47 0	4.0009	4.0009	4.0010	4.0010	4.0010	4.0011	4.0011	4.0012	4.0012	4.0013
47 10	4.0013	4.0013	4.0014	4.0014	4.0015	4.0015	4.0016	4.0016	4.0016	4.0017
47 20	4.0017	4.0018	4.0018	4.0019	4.0019	4.0019	4.0020	4.0020	4.0021	4.0021
47 30	4.0022	4.0022	4.0023	4.0023	4.0023	4.0024	4.0024	4.0025	4.0025	4.0026
47 40	4.0026	4.0026	4.0027	4.0027	4.0028	4.0028	4.0029	4.0029	4.0029	4.0030
47 50	4.0030	4.0031	4.0031	4.0032	4.0032	4.0032	4.0033	4.0033	4.0034	4.0034
2 48 0	4.0035	4.0035	4.0035	4.0036	4.0036	4.0037	4.0037	4.0038	4.0038	4.0038
48 10	4.0039	4.0039	4.0040	4.0040	4.0041	4.0041	4.0041	4.0042	4.0042	4.0043
48 20	4.0043	4.0044	4.0044	4.0045	4.0045	4.0045	4.0046	4.0046	4.0047	4.0047
48 30	4.0048	4.0048	4.0048	4.0049	4.0049	4.0050	4.0050	4.0051	4.0051	4.0051
48 40	4.0052	4.0052	4.0053	4.0053	4.0054	4.0054	4.0054	4.0055	4.0055	4.0056
48 50	4.0056	4.0057	4.0057	4.0057	4.0058	4.0058	4.0059	4.0059	4.0060	4.0060
2 49 0	4.0060	4.0061	4.0061	4.0062	4.0062	4.0063	4.0063	4.0063	4.0064	4.0064
49 10	4.0065	4.0065	4.0066	4.0066	4.0066	4.0067	4.0067	4.0068	4.0068	4.0069
49 20	4.0069	4.0069	4.0070	4.0070	4.0071	4.0071	4.0072	4.0072	4.0072	4.0073
49 30	4.0073	4.0074	4.0074	4.0074	4.0075	4.0075	4.0076	4.0076	4.0077	4.0077
49 40	4.0077	4.0078	4.0078	4.0079	4.0079	4.0080	4.0080	4.0080	4.0081	4.0081
49 50	4.0082	4.0082	4.0083	4.0083	4.0083	4.0084	4.0084	4.0085	4.0085	4.0086

TABLE IX.

LOGARITHMS OF SMALL ARCS IN SPACE OR TIME.										
Arc.	0″	1″	2″	3″	4″	5″	6″	7″	8″	9″
2ʰ·50ᵐ· 0ˢ·	4.0086	4.0086	4.0087	4.0087	4.0088	4.0088	4.0089	4.0089	4.0089	4.0090
50 10	4.0090	4.0091	4.0091	4.0092	4.0092	4.0092	4.0093	4.0093	4.0094	4.0094
50 20	4.0095	4.0095	4.0095	4.0096	4.0096	4.0097	4.0097	4.0097	4.0098	4.0098
50 30	4.0099	4.0099	4.0100	4.0100	4.0100	4.0101	4.0101	4.0102	4.0102	4.0103
50 40	4.0103	4.0103	4.0104	4.0104	4.0105	4.0105	4.0106	4.0106	4.0106	4.0107
50 50	4.0107	4.0108	4.0108	4.0109	4.0109	4.0109	4.0110	4.0110	4.0111	4.0111
2 51 0	4.0111	4.0112	4.0112	4.0113	4.0113	4.0114	4.0114	4.0114	4.0115	4.0115
51 10	4.0116	4.0116	4.0117	4.0117	4.0117	4.0118	4.0118	4.0119	4.0119	4.0120
51 20	4.0120	4.0120	4.0121	4.0121	4.0122	4.0122	4.0122	4.0123	4.0123	4.0124
51 30	4.0124	4.0125	4.0125	4.0125	4.0126	4.0126	4.0127	4.0127	4.0128	4.0128
51 40	4.0128	4.0129	4.0129	4.0130	4.0130	4.0130	4.0131	4.0131	4.0132	4.0132
51 50	4.0133	4.0133	4.0133	4.0134	4.0134	4.0135	4.0135	4.0136	4.0136	4.0136
2 52 0	4.0137	4.0137	4.0138	4.0138	4.0138	4.0139	4.0139	4.0140	4.0140	4.0141
52 10	4.0141	4.0141	4.0142	4.0142	4.0143	4.0143	4.0144	4.0144	4.0144	4.0145
52 20	4.0145	4.0146	4.0146	4.0146	4.0147	4.0147	4.0148	4.0148	4.0149	4.0149
52 30	4.0149	4.0150	4.0150	4.0151	4.0151	4.0152	4.0152	4.0153	4.0153	4.0153
52 40	4.0154	4.0154	4.0154	4.0155	4.0155	4.0156	4.0156	4.0157	4.0157	4.0157
52 50	4.0158	4.0158	4.0159	4.0159	4.0159	4.0160	4.0160	4.0161	4.0161	4.0162
2 53 0	4.0162	4.0162	4.0163	4.0163	4.0164	4.0164	4.0164	4.0165	4.0165	4.0166
53 10	4.0166	4.0167	4.0167	4.0167	4.0168	4.0168	4.0169	4.0169	4.0169	4.0170
53 20	4.0170	4.0171	4.0171	4.0172	4.0172	4.0172	4.0173	4.0173	4.0174	4.0174
53 30	4.0175	4.0175	4.0175	4.0176	4.0176	4.0177	4.0177	4.0177	4.0178	4.0178
53 40	4.0179	4.0179	4.0180	4.0180	4.0180	4.0181	4.0181	4.0182	4.0182	4.0182
53 50	4.0183	4.0183	4.0184	4.0184	4.0185	4.0185	4.0185	4.0186	4.0186	4.0187
2 54 0	4.0187	4.0187	4.0188	4.0188	4.0189	4.0189	4.0190	4.0190	4.0190	4.0191
54 10	4.0191	4.0192	4.0192	4.0192	4.0193	4.0193	4.0194	4.0194	4.0194	4.0195
54 20	4.0195	4.0196	4.0196	4.0197	4.0197	4.0197	4.0198	4.0198	4.0199	4.0199
54 30	4.0199	4.0200	4.0200	4.0201	4.0201	4.0202	4.0202	4.0202	4.0203	4.0203
54 40	4.0204	4.0204	4.0204	4.0205	4.0205	4.0206	4.0206	4.0207	4.0207	4.0207
54 50	4.0208	4.0208	4.0209	4.0209	4.0209	4.0210	4.0210	4.0211	4.0211	4.0211
2 55 0	4.0212	4.0212	4.0213	4.0213	4.0214	4.0214	4.0214	4.0215	4.0215	4.0216
55 10	4.0216	4.0216	4.0217	4.0217	4.0218	4.0218	4.0219	4.0219	4.0219	4.0220
55 20	4.0220	4.0221	4.0221	4.0221	4.0222	4.0222	4.0223	4.0223	4.0223	4.0224
55 30	4.0224	4.0225	4.0225	4.0225	4.0226	4.0226	4.0227	4.0227	4.0228	4.0228
55 40	4.0228	4.0229	4.0229	4.0230	4.0230	4.0230	4.0231	4.0231	4.0232	4.0232
55 50	4.0233	4.0233	4.0233	4.0234	4.0234	4.0235	4.0235	4.0235	4.0236	4.0236
2 56 0	4.0237	4.0237	4.0237	4.0238	4.0238	4.0239	4.0239	4.0240	4.0240	4.0240
56 10	4.0241	4.0241	4.0242	4.0242	4.0242	4.0243	4.0243	4.0244	4.0244	4.0244
56 20	4.0245	4.0245	4.0246	4.0246	4.0246	4.0247	4.0247	4.0248	4.0248	4.0249
56 30	4.0249	4.0249	4.0250	4.0250	4.0251	4.0251	4.0251	4.0252	4.0252	4.0253
56 40	4.0253	4.0253	4.0254	4.0254	4.0255	4.0255	4.0256	4.0256	4.0256	4.0257
56 50	4.0257	4.0258	4.0258	4.0258	4.0259	4.0259	4.0260	4.0260	4.0260	4.0261
2 57 0	4.0261	4.0262	4.0262	4.0262	4.0263	4.0263	4.0264	4.0264	4.0265	4.0265
57 10	4.0265	4.0266	4.0266	4.0267	4.0267	4.0267	4.0268	4.0268	4.0269	4.0269
57 20	4.0269	4.0270	4.0270	4.0271	4.0271	4.0271	4.0272	4.0272	4.0273	4.0273
57 30	4.0273	4.0274	4.0274	4.0275	4.0275	4.0276	4.0276	4.0276	4.0277	4.0277
57 40	4.0278	4.0278	4.0278	4.0279	4.0279	4.0280	4.0280	4.0280	4.0281	4.0281
57 50	4.0282	4.0282	4.0282	4.0283	4.0283	4.0284	4.0284	4.0284	4.0285	4.0285
2 58 0	4.0286	4.0286	4.0287	4.0287	4.0287	4.0288	4.0288	4.0289	4.0289	4.0289
58 10	4.0290	4.0290	4.0291	4.0291	4.0291	4.0292	4.0292	4.0293	4.0293	4.0293
58 20	4.0294	4.0294	4.0295	4.0295	4.0295	4.0296	4.0296	4.0297	4.0297	4.0297
58 30	4.0298	4.0298	4.0299	4.0299	4.0300	4.0300	4.0300	4.0301	4.0301	4.0302
58 40	4.0302	4.0302	4.0303	4.0303	4.0304	4.0304	4.0304	4.0305	4.0305	4.0306
58 50	4.0306	4.0306	4.0307	4.0307	4.0308	4.0308	4.0308	4.0309	4.0309	4.0310
2 59 0	4.0310	4.0310	4.0311	4.0311	4.0312	4.0312	4.0312	4.0313	4.0313	4.0314
59 10	4.0314	4.0314	4.0315	4.0315	4.0316	4.0316	4.0317	4.0317	4.0317	4.0318
59 20	4.0318	4.0319	4.0319	4.0319	4.0320	4.0320	4.0321	4.0321	4.0321	4.0322
59 30	4.0322	4.0323	4.0323	4.0323	4.0324	4.0324	4.0325	4.0325	4.0325	4.0326
59 40	4.0326	4.0327	4.0327	4.0327	4.0328	4.0328	4.0329	4.0329	4.0329	4.0330
59 50	4.0330	4.0331	4.0331	4.0331	4.0332	4.0332	4.0333	4.0333	4.0333	4.0334

TABLE X.

Difference of the Proportional Logarithms in the Ephemeris.

Approximate Interval		2	4	6	8	10	12	14	16	18	20	22	24	26	28	30	32	34	36	38	40	42	44	46	48	50	52
h m	h m	s	s	s	s	s	s	s	s	s	s	s	s	s	s	s	s	s	s	s	s	s	s	s	s	s	s
0 0	3 0	0	0	0	0	0	0	0	0	0	0	0	0	0	0	0	0	0	0	0	0	0	0	0	0	0	0
0 10	2 50	0	0	0	1	1	1	1	1	1	1	1	2	2	2	2	2	2	2	2	3	3	3	3	3	3	3
0 20	2 40	0	1	1	1	1	2	2	2	2	2	3	3	3	3	4	4	4	4	5	5	5	5	6	6	6	6
0 30	2 30	0	1	1	2	2	2	2	3	3	3	4	4	5	5	6	6	6	6	7	7	7	8	8	8	9	9
0 40	2 20	0	1	1	2	2	3	3	3	4	4	5	5	6	6	7	7	8	8	9	9	10	10	10	11	11	
0 50	2 10	1	1	2	2	3	3	4	4	5	5	6	6	7	7	8	8	9	9	10	10	11	12	12	13	13	
1 0	2 0	1	1	2	2	3	3	4	4	5	6	6	7	7	8	8	9	9	10	10	11	12	12	13	13	14	14
1 10	1 50	1	1	2	2	3	4	4	5	5	6	6	7	8	8	9	9	10	11	11	12	12	13	14	14	15	15
1 20	1 40	1	1	2	3	3	4	4	5	6	6	7	7	8	9	9	10	10	11	12	12	13	14	14	15	15	16
1 30	1 30	1	1	2	3	3	4	4	5	6	6	7	8	8	9	9	10	11	11	12	12	13	14	14	15	16	16

Difference of the Proportional Logarithms in the Ephemeris.

Approximate Interval		54	56	58	60	62	64	66	68	70	72	74	76	78	80	82	84	86	88	90	92	94	96	98	100	102
h m	h m	s	s	s	s	s	s	s	s	s	s	s	s	s	s	s	s	s	s	s	s	s	s	s	s	s
0 0	3 0	0	0	0	0	0	0	0	0	0	0	0	0	0	0	0	0	0	0	0	0	0	0	0	0	0
0 10	2 50	4	4	4	4	4	4	4	4	5	5	5	5	5	5	6	6	6	6	6	6	6	6	7		7
0 20	2 40	7	7	7	7	8	8	8	9	9	9	9	10	10	10	10	11	11	11	11	12	12	12	12		13
0 30	2 30	9	10	10	10	11	11	12	12	12	13	13	13	14	14	14	14	15	15	16	16	16	17	17	17	18
0 40	2 20	12	12	13	13	13	14	14	15	15	16	16	16	17	17	18	18	19	19	19	20	20	21	21	22	22
0 50	2 10	14	14	15	15	16	16	16	17	17	18	19	19	20	20	21	21	22	22	22	23	23	24	24	25	26
1 0	2 0	15	16	16	17	17	18	18	19	19	20	21	21	22	22	23	23	24	24	25	25	26	27	27	28	28
1 10	1 50	16	17	17	18	18	19	19	20	21	21	22	22	23	24	24	25	25	26	27	27	28	28	29	30	30
1 20	1 40	17	17	18	19	19	20	20	21	21	22	23	23	24	25	25	26	26	27	28	28	29	29	30	31	31
1 30	1 30	17	18	18	19	19	20	21	21	22	23	23	24	24	25	25	26	27	27	28	29	29	30	31	31	32

Difference of the Proportional Logarithms in the Ephemeris.

Approximate Interval		104	106	108	110	112	114	116	118	120	122	124	126	128	130	132	134	136	138
h m	h m	s	s	s	s	s	s	s	s	s	s	s	s	s	s	s	s	s	s
0 0	3 0	0	0	0	0	0	0	0	0	0	0	0	0	0	0	0	0	0	0
0 10	2 50	7	7	7	7	7	8	8	8	8	8	8	8	9	9	9	9	9	9
0 20	2 40	13	13	13	14	14	14	14	15	15	15	15	15	16	16	16	16	17	17
0 30	2 30	18	18	19	19	19	20	20	20	21	21	21	22	22	22	23	23	24	24
0 40	2 20	22	23	23	24	24	25	25	25	26	26	27	27	28	28	28	29	29	30
0 50	2 10	26	26	27	27	28	29	29	29	30	30	31	31	32	32	33	33	34	34
1 0	2 0	29	29	30	30	31	31	32	33	33	34	34	35	35	36	37	37	38	39
1 10	1 50	31	31	32	32	33	34	34	35	35	36	37	37	38	33	39	40	40	41
1 20	1 40	32	33	33	34	34	35	35	36	37	38	38	39	39	40	41	41	42	42
1 30	1 30	32	33	34	34	35	35	36	36	37	38	39	39	40	40	41	42	42	43

The Correction is to be *added* to the approximate Greenwich Time when the Proportional Logarithms in the Ephemeris are *decreasing*, and *subtracted* when they are *increasing*.

TABLE XI.

LOG. N FOR DISTANCES FROM THE SUN.

1855.	0ʰ	3ʰ	6ʰ	9ʰ	12ʰ	15ʰ	18ʰ	21ʰ	1855.	0ʰ	3ʰ	6ʰ	9ʰ	12ʰ	15ʰ	18ʰ	21ʰ
Jan. 7	−0.76	.78	.79	.80	.81	.82	.83	.84	Apr. 10	+0.66	.69	.72	.74	.77	.79	.81	.83
8	0.85	.86	.87	.87	.88	.89	.89	.90	11	0.85	0.86	0 88	0.89	0.91	0.92	0.93	0.95
9	0.90	.91	.91	.91	.92	.92	.92	.92	12	0.96	0.97	0.98	0.99	1.00	1.01	1.02	1.02
10	0.92	.92	.92	.92	.92	.92	.92	.92	13	+1.03	1.04	1.04	1.05	1.06	1.06	1.07	1.07
11	0.92	.91	.91	.91	.90	.90	.89	.89	18	−0.80	0.78	0.77	0.76	0.75	0.73	0.71	0.69
12	−0.88	.87	.86	.85	.84	.83	.82	.81	19	−0.67	.65	.63	.60	.58	.55	.52	.48
13	0.80	.78	.77	.75	.73	.71	.69	.66	20	−0.46	.41	.36	.31	.25	.18	.10	.00
14	0.63	.60	.56	.52	.48	.42	.36	.28	22	+0.12	.20	.26	.32	.37	.41	.45	.49
20	0.71	.74	.76	.79	.81	.83	.84	.85	23	0.52	.55	.58	.60	.63	.65	.67	.69
21	0.87	.88	.89	.89	.90	.91	.91	.92	24	0.71	.73	.75	.76	.78	.79	.80	.82
22	−0.92	.93	.93	.93	.93	.93	.93	.93	25	+0.83	.84	.85	.86	.87	.88	.89	.90
23	0.93	.93	.93	.92	.92	.91	.91	.90	26	0.91	.92	.93	.93	.94	.95	0.95	0.96
24	0.90	.89	.89	.88	.87	.86	.85	.84	27	0.97	0.97	0.98	.98	.99	.99	0.99	1.00
25	0.83	.82	.81	.80	.78	.77	.75	.74	May 6	9.77	9.96	0.10	.20	.28	.35	0.40	0.46
26	0.72	.70	.68	.66	.04	.61	.59	.56	7	0.50	0.54	0.58	.61	.04	.67	.70	.72
27	−0.53	.49	.45	.41	.37	.31	.25	.18	8	+0.75	.77	.79	.80	.82	.84	.85	.87
Feb. 6	0.88	.88	.88	.89	.89	.89	.89	.89	9	0 88	.89	.90	.92	.93	.94	0.94	0.95
7	0.89	.89	.89	.89	.89	.88	.88	.88	10	0.96	0.97	0.97	0.98	0.99	0.99	0.99	1.00
8	0.87	.87	.86	.86	.85	.84	.83	.82	11	1.00	1.01	1.01	1.01	1.01	1.01	1.01	1 01
9	0.82	.80	.79	.78	.77	.75	.74	.72	12	1.01	1.01	1.01	1.01	1.01	1.00	1.00	1.00
10	−0.70	.68	.66	.64	.61	.58	.55	.52	13	+0.99	0.09	0.98	0.97	0.97	0.96	0.95	0.94
11	−0.48	.43	.38	.32	.25	.16	.05	9.89	18	−0.46	.41	.35	.29	.22	.13	0.03	9.89
13	+0.43	.49	.55	.60	.64	.69	.73	.76	20	+0.23	.29	.34	.39	.43	.47	0.51	0.54
14	+0.80	.83	.87	.90	.93	.96	.99	1.02	21	0.57	.59	.62	.64	.66	.68	.70	.72
18	−0.78	.80	.82	.84	.85	.86	.87	.88	22	0.73	.75	.76	.78	.79	.81	.82	.83
19	−0.89	.89	.90	.90	.90	.90	.90	.90	23	+0.84	.85	.86	.87	.88	.88	.89	.90
20	0.90	.89	.89	.89	.88	.88	.87	.86	24	0.91	.91	.92	.93	.93	.94	.94	.94
21	0.86	.85	.84	.83	.82	.81	.79	.78	25	0.95	.95	.96	.96	.96	.96	.96	.97
22	0.77	.75	.74	.72	.70	.69	.67	.64	26	0.97	.97	.97	.97	.97	.97	.97	.96
23	0.62	.60	.57	.54	.51	.48	.44	.40	27	0.96	.96	.96	.95	.95	.94	.94	.98
26	+0.30	.36	.41	.45	.49	.53	.56	.60	June 3	+0.05	.18	.28	.35	.42	.47	.52	.56
Mar. 8	−0.83	.82	.81	.80	.79	.78	.77	.76	4	0.60	.63	.66	.69	.72	.74	.76	.78
9	0.74	.73	.71	.70	.68	.66	.64	.60	5	0.80	.82	.83	.85	.86	.87	.88	.89
10	−0.59	.56	.53	.49	.45	.41	.36	.30	6	0.90	.91	.92	.93	.94	.94	.95	.95
12	+0.91	.06	.17	.26	.34	.40	.46	.51	7	0.96	.96	.97	.97	.97	.97	.98	.98
13	+0.55	.59	.63	.67	.70	.73	.76	.78	8	+0.98	.98	.98	.98	.97	.97	.97	.97
14	0.81	0.83	0.85	0.87	0.89	0.91	0.93	0 95	9	0.96	.96	.95	.95	.94	.94	.93	.92
15	+0.97	0.99	1.00	1.02	1.03	1.05	1.06	1.08	10	0.91	.91	.90	.88	.87	.86	.85	.83
19	−0.78	0.80	0.82	0.84	0.84	0.85	0 85	0.86	11	0.82	.80	.78	.76	.73	.71	.68	.64
20	−0.86	.86	.85	.85	.84	.84	.83	.82	17	0.18	.26	.32	.38	.43	.47	.51	.54
21	−0.82	.81	.80	.78	.77	.76	.74	.73	18	+0.57	.60	.63	.65	.67	.69	.71	.73
22	0.71	.69	.67	.65	.63	.61	.58	.55	19	0.74	.76	.77	.79	.80	.81	.82	.83
23	−0.53	.49	.46	.42	.38	.33	.27	.21	20	0.84	.85	.86	.87	.87	.88	.89	.89
25	+0.84	.98	.08	.16	.23	.29	.35	.39	21	0.90	.90	.91	.91	.92	.92	.93	.93
26	0.44	.48	.51	.54	.57	.60	.63	.65	22	0.93	.93	.92	.94	.94	.94	.94	.94
27	+0.67	.69	.71	.73	75	.77	.79	.80	23	+0.94	.94	.94	.93	.93	.93	.93	.92
28	+0.82	.83	.85	.86	.87	.88	.90	.91	24	0.92	.91	.91	.90	.90	.89	.88	.88
Apr. 6	−0.67	.65	.63	.61	.59	.55	.52	.48	25	0.87	.86	.85	.83	.82	.81	.79	.78
7	−0.44	.40	.35	.29	.22	.13	.03	.88	July 3	0.82	.84	.85	.87	.88	.89	.90	.91
9	+0.31	.37	.43	.48	.52	.56	.60	.63	4	0.92	.92	.93	.93	.93	.94	.95	.95

TABLE XI.

LOG. N FOR DISTANCES FROM THE SUN.

1855.	0h.	3h.	6h.	9h.	12h.	15h.	18h.	21h.	1855.	0h.	3h.	6h.	9h.	12h.	15h.	18h.	21h.
July 5	+0.95	.95	.96	.96	.96	.95	.95	.95	Oct. 3	−0.40	.44	.48	.52	.55	.58	.61	.63
6	0.95	.95	.94	.94	.94	.93	.93	.92	4	0.66	.68	.70	.72	.74	.76	.77	.79
7	0.91	.91	.90	.89	.88	.87	.86	.85	5	0.80	.82	.83	.84	.86	.87	.88	.89
8	0.84	.83	.81	.80	.78	.76	.75	.73	6	0.90	.91	.92	0.93	0.94	0.95	0.96	0.96
9	0.70	.68	.66	.63	.60	0.57	0.53	0.49	7	−0.97	.98	.98	0.99	1.00	1.00	1.01	1.02
10	+0.44	.39	.38	.25	.16	0.04	9.86	9.54	13	+0.84	.83	.82	0.81	0.80	0.78	0.77	0.76
16	0.47	.52	.56	.60	.63	0.66	0.68	0.70	14	0.74	.72	.71	.69	.67	.64	.62	.59
17	0.72	.74	.76	.77	.79	.80	.81	.82	15	+0.56	.53	.50	.46	.42	.37	.31	.25
18	0.83	.84	.85	.86	.86	.87	.88	.88	17	−9.98	.10	.20	.28	.34	.40	.45	.50
19	0.89	.89	.90	.90	.90	.91	.91	.91	18	0.54	.58	.61	.64	.67	.70	.73	.75
20	+0.91	.91	.91	.91	.91	.91	.91	.91	19	−0.77	.79	.81	.83	.85	.86	.88	.89
21	0.91	.91	.90	.90	.90	.89	.89	.88	20	0.91	.92	.93	.94	.96	.97	.98	.99
22	0.88	.87	.86	.85	.85	.84	.83	.82	30	9.99	.10	.18	.25	.31	.36	.41	.45
23	0.80	.79	.78	.76	.75	.73	.71	.69	31	0.49	.52	.55	.58	.61	.63	.65	.68
24	0.67	.64	.62	.59	.55	.52	.47	.42	Nov. 1	0.70	.72	.73	.75	.77	.78	.79	.81
Aug. 1	+0.91	.92	.92	.93	.93	.93	.94	.94	2	−0.82	.83	.84	.85	.86	.87	.88	.89
2	0.94	.94	.94	.94	.93	.93	.93	.92	3	0.90	.91	.91	.92	.93	.93	.94	.94
3	0.92	.91	.91	.90	.90	.89	.88	.87	4	0.95	.95	.96	.96	.96	.97	.97	.97
4	0.86	.85	.86	.83	.82	.80	.79	.77	5	0.98	.98	.98	.98	.98	.98	.98	.98
5	0.76	.74	.72	.70	.68	.66	.63	.61	6	−0.98	.98	.98	.98	.98	.97	.97	.97
6	+0.58	.55	.51	.47	.43	.38	.83	.27	12	+0.53	.48	.43	.37	.31	.23	.14	.02
8	−9.93	.07	.17	.26	.33	.39	.45	.50	14	−0.28	.35	.41	.46	.50	.54	.58	.61
9	−0.54	.58	.62	.66	.69	.72	.75	.78	15	0.64	.67	.76	.72	.74	.77	.79	.80
15	+0.79	.81	.82	.83	.84	.85	.86	.86	16	0.82	.84	.85	.87	.88	.89	.90	.91
16	0.87	.87	.88	.88	.88	.89	.89	.89	17	0.92	.93	.94	.95	.96	0.96	0.97	0.97
17	+0.89	.89	.89	.89	.89	.88	.88	.88	18	−0.98	0.98	0.99	0.99	0.99	1.00	1.00	1.00
18	0.88	.87	.87	.86	.86	.85	.84	.84	19	1.00	1.00	1.00	1.00	1.00	1.00	1.00	0.99
19	0.83	.82	.81	.80	.79	.78	.76	.75	27	0.06	0.16	0.24	0.31	0.36	0.41	.45	.49
20	0.73	.72	.70	.68	.66	.64	.61	.58	28	0.53	.56	.59	.62	.64	.66	.68	.70
21	+0.55	.52	.48	.44	.40	.34	.28	.20	29	0.72	.73	.75	.76	.78	.79	.80	.82
23	−0.17	.27	.35	.42	.48	.53	.58	.60	30	−0.83	.84	.85	.86	.86	.87	.88	.89
30	+0.92	.92	.92	.91	.91	.91	.90	.90	Dec. 1	0.89	.90	.91	.91	.92	.92	.93	.93
31	0.89	.88	.87	.87	.86	.85	.84	.82	2	0.93	.94	.94	.94	.94	.95	.95	.95
Sept. 1	0.81	.70	.78	.77	.75	.74	.72	.70	3	0.95	.95	.95	.95	.95	.95	.94	.94
2	0.67	.65	.63	.60	.57	.54	.51	.47	4	0.94	.94	.93	.93	.93	.92	.92	.91
3	+0.43	.38	.33	.27	.20	.12	.02	.88	5	−0.90	.90	.89	.88	.87	.86	.85	.84
5	−0.24	.31	.36	.41	.46	.50	.53	.57	6	0.82	.81	.79	.77	.75	.73	.70	.67
6	0.60	.63	.65	.68	.70	.73	.75	.77	12	0.38	.44	.49	.54	.58	.62	.65	.68
7	−0.79	.81	.83	.84	.86	.88	.89	.71	13	0.71	.73	.75	.77	.79	.81	.83	.84
13	+0.83	.84	.85	.85	.86	.86	.86	.87	14	0.86	.87	.88	.89	.90	.91	.92	.93
14	+0.87	.87	.87	.86	.86	.86	.86	.85	15	−0.93	.94	.95	.95	.96	.96	.96	.97
15	0.85	.84	.83	.83	.82	.81	.80	.79	16	0.97	.97	.97	.97	.97	.97	.97	.97
16	0.78	.77	.75	.74	.72	.71	.69	.67	17	0.97	.96	.96	.96	.95	.95	.94	.94
17	0.65	.63	.60	.58	.55	0.52	0.48	0.44	18	0.93	.92	.91	.90	.89	.88	.87	.86
18	+0.40	.35	.29	.22	.14	0.03	9.89	9.67	27	0.72	.73	.75	.76	.78	.79	.80	.81
20	−0.39	.44	.49	.54	.58	0.62	0.65	0.69	28	−0.83	.84	.84	.85	.86	.87	.88	.88
21	−0.71	.74	.77	.79	.82	.84	.86	.88	29	0.89	.89	.90	.90	.91	.91	.91	.92
29	+0.77	.75	.74	.72	.70	.68	.65	.63	30	0.92	.92	.92	.92	.93	.93	.93	.93
30	+0.60	0.57	0.54	0.50	.46	.42	.37	.32	31	−0.93	.92	.92	.92	.92	.92	.91	.91
Oct. 2	−9.58	9.81	9.97	0.08	.16	.24	.30	.35									

TABLE XII.

For finding the value of N for correcting lunar distances for the compression of the earth.

TABLE XII. *A.* giving 1st Part of N.

D's Declination.

App. Dist.	0	3	6	9	12	15	18	21	24	27	30
20	−0	3	6	10	13	16	19	22	25	28	31
22	0	3	6	9	12	14	17	20	23	25	28
24	0	3	5	8	11	13	16	18	21	23	25
26	0	2	5	7	10	12	14	17	19	21	23
28	0	2	4	7	9	11	13	15	17	19	21
30	−0	2	4	6	8	10	12	14	16	18	20
32	0	2	4	6	8	9	11	13	15	16	18
34	0	2	4	5	7	9	10	12	14	15	17
36	0	2	3	5	7	8	10	11	13	14	16
38	0	2	3	5	6	8	9	10	12	13	14
40	−0	1	3	4	6	7	8	10	11	12	13
42	0	1	3	4	5	7	8	9	10	11	13
44	0	1	2	4	5	6	7	8	10	11	12
46	0	1	2	3	5	6	7	8	9	10	11
48	0	1	2	3	4	5	6	7	8	9	10
50	−0	1	2	3	4	5	6	7	8	9	10
52	0	1	2	3	4	5	5	6	7	8	9
54	0	1	2	3	3	4	5	6	7	7	8
56	0	1	2	2	3	4	5	5	6	7	8
58	0	1	1	2	3	4	4	5	6	6	7
60	−0	1	1	2	3	3	4	5	5	6	7
62	0	1	1	2	3	3	4	4	5	5	6
64	0	1	1	2	2	3	3	4	4	5	6
66	0	1	1	2	2	3	3	4	4	5	5
68	0	0	1	1	2	2	3	3	4	4	5
70	−0	0	1	1	2	2	3	3	4	4	4
72	0	0	1	1	2	2	2	3	3	3	4
74	0	0	1	1	1	2	2	2	3	3	3
76	0	0	1	1	1	1	2	2	2	3	3
78	0	0	0	0	1	1	1	1	2	2	2
80	−0	0	0	1	1	1	1	1	2	2	2
82	0	0	0	0	1	1	1	1	1	1	2
84	0	0	0	0	0	1	1	1	1	1	1
86	0	0	0	0	0	0	0	1	1	1	1
88	0	0	0	0	0	0	0	0	0	0	0
90	−0	0	0	0	0	0	0	0	0	0	0
92	+0	0	0	0	0	0	0	0	0	0	0
94	0	0	0	0	0	0	0	1	1	1	1
96	0	0	0	0	0	1	1	1	1	1	1
98	0	0	0	0	1	1	1	1	1	1	2
100	+0	0	0	1	1	1	1	1	2	2	2
102	0	0	0	1	1	1	1	2	2	2	2
104	0	0	1	1	1	1	2	2	2	3	3
106	0	0	1	1	1	2	2	2	3	3	3
108	0	0	1	1	2	2	2	3	3	3	4
110	+0	0	1	1	2	2	3	3	3	4	4
112	0	0	1	1	2	2	3	3	4	4	5
114	0	1	1	2	2	3	3	4	4	5	5
116	0	1	1	2	2	3	3	4	4	5	6
118	0	1	1	2	3	3	4	4	5	5	6
120	+0	1	1	2	3	3	4	5	5	6	7
122	0	1	1	2	3	4	4	5	6	6	7
124	0	1	2	2	3	4	5	5	6	7	8
126	0	1	2	3	3	4	5	6	7	7	8
128	0	1	2	3	4	5	5	6	7	8	9
130	+0	1	2	3	4	5	6	7	8	9	10

TABLE XII. *B.* giving 2d Part of N.

*'s Declination.

App. Dist.	0	3	6	9	12	15	18	21	24	27	30
20	+0	3	7	10	14	17	20	24	27	30	33
22	0	3	6	9	13	16	19	22	25	27	30
24	0	3	6	9	12	14	17	20	23	25	28
26	0	3	5	8	11	13	16	18	21	23	26
28	0	3	5	8	10	12	15	17	20	22	24
30	+0	2	5	7	9	12	14	16	18	21	23
32	0	2	4	7	9	11	13	15	17	19	21
34	0	2	4	6	8	11	13	15	16	18	20
36	0	2	4	6	8	10	12	14	16	17	19
38	0	2	4	6	8	10	11	13	15	17	18
40	+0	2	4	6	7	9	11	13	14	16	18
42	0	2	4	5	7	9	10	12	14	15	17
44	0	2	3	5	7	8	10	12	13	15	16
46	0	2	3	5	6	8	10	11	13	14	16
48	0	2	3	5	6	8	9	11	12	14	15
50	+0	2	3	5	6	8	9	11	12	13	15
52	0	2	3	4	6	7	9	10	12	13	14
54	0	1	3	4	6	7	9	10	11	13	14
56	0	1	3	4	6	7	8	10	11	12	14
58	0	1	3	4	6	7	8	10	11	12	13
60	+0	1	3	4	5	7	8	9	11	12	13
62	0	1	3	4	5	7	8	9	10	12	13
64	0	1	3	4	5	7	8	9	10	11	13
66	0	1	3	4	5	6	8	9	10	11	12
68	0	1	3	4	5	6	8	9	10	11	12
70	+0	1	3	4	4	6	7	9	10	11	12
72	0	1	2	4	5	6	7	9	10	11	12
74	0	1	2	4	5	6	7	8	10	11	12
76	0	1	2	4	5	6	7	8	9	11	12
78	0	1	2	4	5	6	7	8	9	11	12
80	+0	1	2	4	5	6	7	8	9	10	11
82	0	1	2	4	5	6	7	8	9	10	11
84	0	1	2	4	5	6	7	8	9	10	11
86	0	1	2	4	5	6	7	8	9	10	11
88	0	1	2	4	5	6	7	8	9	10	11
90	+0	1	2	4	5	6	7	8	9	10	11
92	0	1	2	4	5	6	7	8	9	10	11
94	0	1	2	4	5	6	7	8	9	10	11
96	0	1	2	4	5	6	7	8	9	10	11
98	0	1	2	4	5	6	7	8	9	10	11
100	+0	1	2	4	5	6	7	8	9	10	11
102	0	1	2	4	5	6	7	8	9	11	12
104	0	1	2	4	5	6	7	8	9	11	12
106	0	1	2	4	5	6	7	8	10	11	12
108	0	1	2	4	5	6	7	9	10	11	12
110	+0	1	3	4	5	6	7	9	10	11	12
112	0	1	3	4	5	6	7	9	10	11	12
114	0	1	3	4	5	6	8	9	10	11	12
116	0	1	3	4	5	7	8	9	10	11	13
118	0	1	3	4	5	7	8	9	10	12	13
120	+0	1	3	4	5	7	8	9	11	12	13
122	0	1	3	4	6	7	8	10	11	12	13
124	0	1	3	4	6	7	8	10	11	12	14
126	0	1	3	4	6	7	9	10	11	13	14
128	0	2	3	4	6	7	9	10	12	13	14
130	+0	2	3	5	6	8	9	11	12	13	15

The signs in the 0 column apply to all the numbers in the same line, and are to be used when the declination is *North.* When the declination is *South,* change the sign + to − and − to +.

IMPROVED METHOD

OF FINDING THE

ERROR AND RATE OF A CHRONOMETER

BY EQUAL ALTITUDES.

FROM THE APPENDIX TO THE AMERICAN EPHEMERIS AND NAUTICAL ALMANAC FOR 1856.

METHOD

OF FINDING THE

ERROR AND RATE OF A CHRONOMETER BY EQUAL ALTITUDES.

———————

To regulate a chronometer to Greenwich time, we must determine its error and rate at a place whose longitude is well known. The most accurate method of doing this is by observing the transit of the sun or a star over the meridian. For the navigator, the most simple and accurate substitute for the meridian observation is that of equal altitudes of the same object on each side of the meridian. In the case of a star, the mean of the two chronometer times corresponding to the equal altitudes is the chronometer time of transit; but in the case of the sun, the mean of these times differs somewhat from the time of transit, since, in consequence of the change of the sun's declination between the observations, the equal altitudes do not occur at equal intervals before and after the transit.

The small correction necessary, when the sun is observed, to reduce the mean of the times to the time of transit, is called the *Equation of Equal Altitudes.*

The method of computing this equation given below is based upon that first given by GAUSS (*Monatliche Correspondenz*, Vol. XXIII.). We do not, however, follow him in using the double daily change of declination, or difference between the sun's declination on the noon preceding and the noon following that of the observation; but prefer to use the hourly difference, because this may be obtained directly from the American Ephemeris, and is at the same time even more accurate. We also extend our table so as to meet the case where one altitude is taken in the afternoon and the corresponding equal altitude on the following morning; in which case, the equation is computed for apparent midnight.*

———————

* It should be observed, as a caution to navigators, that the rule for computing the equation for midnight is sometimes inaccurately, or incompletely, stated in works on navigation or astronomy. The rule in Lieut. RAPER's *Practice of Navigation* is wholly erroneous. GALBRAITH's rule (*Mathematical and Astronomical Tables*) is incomplete, in not noticing the case where the elapsed time is less than 12^h. His rule for computing the equation for noon is similarly defective, in not noticing the case where the elapsed time is greater than 12^h. In Professor INMAN's rule there is a slight inaccuracy introduced, by taking the equation of time for mean, instead of apparent noon or midnight; and in all the books,

I. EQUAL ALTITUDES OF THE SUN, MORNING AND EVENING.

THE OBSERVATION.

On shore, at a place whose longitude is *accurately* known, and whose latitude is *approximately* known, observe with an artificial horizon the same altitude both in the morning and in the afternoon, as near the prime vertical as convenient after the altitude is more than 10°, noting the times by the chronometer. In low latitudes, however, the method of equal altitudes will often give very accurate results, even when the observations are quite near to the meridian. In general, a sufficiently accurate result may be obtained if the observations are taken when the sun's change of altitude is not less than 10″ in 0ˢ.5, or when the change in the double altitude taken with the artificial horizon is not less than 20″ in 0ˢ.5.

It is most convenient, as well as conducive to accuracy, to take the observation in the following manner. In the morning, bring the lower limb of the sun, reflected from the sextant-mirrors, and the upper limb of that reflected from the mercury, into approximate contact; move the 0 of the vernier forward (say from 10′ to 20′), and set it on a division of the limb; the images will be *overlapped* and will be *separating*; wait for the instant of contact; note it by chronometer, and immediately set the vernier on the next division of the limb, that is, 10′ in advance; note the instant of contact again, and proceed in the same manner for as many observations as are thought necessary. If the sun rises too rapidly, let the intervals on the limb be 20′. Find (roughly) the time when the sun will be at the same altitude in the afternoon, and just before that time set the vernier on the last altitude noted in the morning (of course using the same sextant); the images of the sun will be *separated*, but will be *approaching;* wait for the instant of contact; note it by chronometer; set the vernier *back* to the next division of the limb (10′ or 20′, as the case may be); note the contact again, and so proceed till all the A. M. altitudes have been again noted as P. M. altitudes.

THE COMPUTATION.

Take the mean of the A. M. times and call it the *A. M. Chronometer Time.* The mean of the P. M. times call the *P. M. Chronometer Time.* If, instead of noting the times by the chronometer, a watch is used (compared with the chronometer both before and after each observation), it will generally be found necessary to make an allowance for its gain or loss on the chronometer, so as to obtain the exact difference between the watch and chronometer at the instant of observation. This difference being applied to the mean of the watch times, we have the mean chronometer time the same as would have been found by employing the chronometer directly.

---

the methods given of taking out the sun's change of declination, whether for 48ʰ or for 24ʰ, are not as accurate as they should be.

A perfectly accurate rule, with a special table, for the midnight correction, is given in SCHUMACHER's *Hülfstafeln* (Ed. by WARNSTORFF). It requires, however, one logarithm more than our method in the text, and is otherwise not so simple.

EQUAL ALTITUDES.

The half sum of the A. M. and P. M. Chronometer Times is the *Middle Chronom-eter Time*, their difference is the *Elapsed Time;* observing that when the A. M. time is before 12^h by chronometer, while the P. M. time is after 12^h, the latter must be supposed to be increased by 12^h in finding this half sum and difference.

Take from the Nautical Almanac the sun's declination, the hourly difference of declination, and the equation of time, reducing each to the instant of local apparent noon by applying the changes for the longitude.

Mark *north* latitude and *north* declination +

    " *south* latitude and *south* declination —

    " hourly diff. of decl. when *towards north* +

    " hourly diff. of decl. when *towards south* —.

Enter Table I. with the elapsed time, and take out log. A and log. B, prefixing to each its proper sign given in the table at the head of the page.

To log. A add the log. of the hourly diff., Table II., and the log. tangent of the latitude (Bowditch, Table XXVII.). Prefix to each log. the sign of the quantity it represents and to their sum the sign which results from the algebraic combination of the three signs.* This sum is the log. (Table II.) of the number of seconds of time in the *first part* of equation of equal altitudes, to be marked + or — like its log.

To log. B. add the log. of the hourly diff. and the log. tangent of the declination, marking the signs as before. The sum is the log. of the *second part* of the equation of equal altitudes, to be marked + or — like its log.

Apply the two parts of the equation, according to their signs, to the Middle Chro-nometer Time; the result is the *Chronometer Time of Apparent Noon.*

To this apply the equation of time (adding, when the equation of time is additive to mean time, otherwise subtracting); the result is the *Chronometer Time of Mean Noon*, which, if the chronometer is regulated to local time, will be 12^h 0^m 0^s when the chronometer is right; more than 12^h when fast, less than 12^h when slow.

If the chronometer is regulated to Greenwich time, apply the longitude (in time) to the chronometer time of mean noon (subtracting in west, adding in east); the result will be more or less than 12^h, according as the chronometer is fast or slow.

Repeat this process on a subsequent day. The difference between the chronome-ter errors on the two days, divided by the number of days in the interval, is the *daily rate* of the chronometer, *gaining* or *losing* according as the chronometer goes too fast or too slow.

Example 1.

May 3d, 1856. At the United States Naval Academy, Lat. 38° 59′ N., Long. 5^h 5^m $55^s.1$ W., suppose the following observations of equal altitudes to be taken with an artificial horizon. Required the error of the chronometer on Greenwich time at noon of that day?

---

* The algebraic rule being, that, when there is an *odd* number of factors with the sign minus, the result must have the sign minus, otherwise the sign plus. In the present application of this rule, when there is either *one* or *three* of the logs. marked —, their sum must be marked —; otherwise +.

A. M. Comparisons.					P. M. Comparisons.		
	h. m. s.					h. m. s.	
Chronom.	12 52 0.0	A. M., watch gains 1ˢ.5 in 28ᵐ·			Chronom.	8 37 0.0	
Watch	7 45 8.0	Interval to obs. 17ᵐ.5			Watch	3 30 31.3	
Diff.	5 6 52.0	28ᵐ· : 17ᵐ.5 = 1ˢ.5 : 0ˢ.9			Diff.	5 6 28.7	
Chronom.	1 20 0.0				Chronom.	9 11 0.0	
Watch	8 13 9.5	P. M., watch gains 2ˢ.2 in 34ᵐ·			Watch	4 4 33.5	
Diff.	5 6 50.5	Interval to obs. 21ᵐ.			Diff.	5 6 26.5	
		34ᵐ· : 21ᵐ· = 2ˢ.2 : 1ˢ.4					

Watch A. M.		2 ☉ Art. Hor.			Watch P. M.		
	h. m. s.		° ′			h. m. s.	
	8 2 9.		65 50			3 52 10.7	
	8 2 35.5		66 0			3 51 44.0	
	8 3 0.5		66 10			3 51 18.5	
Mean	8 2 35.0				Mean	3 51 44.4	
Comparison	5 6 51.1		m. s.	s.	Comparison	5 6 27.3	
A. M. Chro. Time	1 9 26.1	(Eq. T.)	−3 18.11	0.258	P. M. Chro. Time	8 58 11.7	
P. M. Chro. Time	8 58 11.7		1.32	5.1	A. M. Chro. Time	1 9 26.1	
	2)10 7 37.8	Eq. T.	−3 19.43	1.32	Elapsed Time	7 48 45.6	
Middle Chro. T.	5 3 48.9						
Equat. of Eq. Alts.	−8.8	(D.) +15 48 50.5		(H. D.) +43.82	Decrease in 24.0 = 0.66		
Chro. T. App. N.	5 3 40.1	3 42.8		−0.14	Decrease in 5.1 = 0.14		
Equat. of Time	+3 19.4	D. +15 52 33.3		H. D. +43.68			
Chro. T. Mean N.	5 6 59.5			5.1			
Longitude	5 5 55.1 W.			222.8			
Chro. *fast*	1 4.4						

log.A.Tab.I. −9.4846   log. B. Tab. I. +9.2011
H D.+43″.68 log. Tab. II. +1.6403   . . . . +1.6403
Lat. + 38° 59′ log. tan. +9.9081   D. + 15° 53′ log. tan. +9.4542
1st Pt. Eq. −10ˢ.79   log. −1.0330   2d Pt.Eq. +1ˢ.98 log. +0.2956

By similar observations on May 15th, suppose the chronometer is found to be fast 12ˢ.5; we have

	m. s.
May 3d, fast	1 4.4
May 15th, fast	12.5
Loses in 12 days	51.9
Daily rate	4.33 losing.

II. EQUAL ALTITUDES OF THE SUN, EVENING AND MORNING.

THE OBSERVATION.

Take a set of altitudes, in the manner already explained, in the afternoon of one day, and the same altitudes in reverse order on the morning of the next, noting the times by the chronometer, or by a watch compared with it.

THE COMPUTATION.

The half sum of the P. M. and A. M. Chronometer Times is the *Middle Chronometer Time;* their difference is the *Elapsed Time;* observing that when the P. M. time is before 12ʰ by chronometer, while the A. M. time is after 12ʰ, the latter must be supposed to be increased by 12ʰ in finding this half sum and this difference.

Take from the Nautical Almanac the sun's declination, the hourly difference of declination, and the equation of time, reducing them each to the instant of local *apparent midnight.*

68

EQUAL ALTITUDES.

Mark the sign of each quantity as before, and compute the two parts of the equation of equal altitudes precisely as in the preceding case, observing to mark the signs of log. A and log. B as given in the table for midnight.

Apply the two parts of the equation to the middle chronometer time, according to their signs; the result is the *Chronometer Time of Apparent Midnight.*

To this apply the equation of time (adding, when the equation of time is additive to mean time, otherwise subtracting); the result is the *Chronometer Time of Mean Midnight,* which, if the chronometer is regulated to local time, will be 12^h· 0^m· 0^s· when the chronometer is right; more than 12^h· when fast; less than 12^h· when slow.

If the chronometer is regulated to Greenwich time, apply the longitude, in time, to the chronometer time of mean midnight (subtracting in west, adding in east); the result will be more or less than 12^h· (or 24^h·) according as the chronometer is fast or slow.

A repetition of this process at a subsequent day will give another error, whence the rate will be found as before. Or the rate may be found by comparing the results of an A. M. — P. M., and a P. M. — A. M. observation, remembering that the interval elapsed between two such observations is equal to the difference between the two dates *plus* or *minus* half a day.

EXAMPLE 2.

May 3d, 1856, Lat. 43° 21' S., Long. 9^h· 50^m· 8^s· E., suppose the altitude of the sun to be observed in the afternoon and the same altitude again on the morning of the 4th, as below. Required the error of the chronometer on Greenwich time at midnight of the 3d ?

Chronom., P. M.	2 ☉Art. Hor.	Chronom., A. M.
6^h· 54^m· 10^s·3	38° 0'	9^h· 9^m· 17^s·5

The A. M. time must be called 21^h· 9^m· 17^s·5. The Greenwich time of midnight for which the declination, &c. must be found, is May 3^d· 2^h· 9^m· 52^s· (= 3^d· 2^h·.16.).

	h. m. s.			m s.	s.		h. m. s.
P. M. Chro. T.	6 54 10.3		(Eq. T.)	−3 18.11	0.258	A. M. Chro. T.	21 9 17.5
A. M. Chro. T.	21 9 17.5			0.56	2.15	P. M. Chro. T.	6 54 10.3
	2)28 3 27.8		Eq. T.	−3 18.67	0.56	Elapsed T.	14 15 7.2
Middle Chro. T.	14 1 43.9						
Eq. of Eq. Alts.	−22.3						
Chro.T.App.Midn.	14 1 21.6		(D.) +15 48 50.5	(H. D.) +43.82	Decrease in 24.0 = 0.66	
Eq. of Time	+3 18.7		1 34.1	−0.06	Decrease in 2.16 = 0.06	
Chro.T.M'n Midn.	14 4 40.3		D. +15 50 24.6	H. D. +43.76		
Longitude	9 50 8.0 E.			2.15		
	23 54 48.3			94.1		
	24 0 0.0						
Chronom. *slow*	5 11.7		log.A.Tab.I. +9.6958	log. B. Tab. I. −9.1586	

H.D.+43″.76 log. Tab. II. +1.6411 . . . . +1.6411
Lat. −43° 21' log. tan. −9.9750 D + 15° 50' log. tan. +9.4527
1st Pt. Eq. −20″.51 log. −1.3119 2dPt.Eq. − 1^s.79 log. −0.2524

By an A. M. — P. M. observation on May 20th, suppose this chronometer is found to be slow 8^m· 14^s.6; we have

	d. h.		m. s.
May	3 12	slow	5 11.7
May	20 0	slow	8 14.6
Loses in 16^d·.5			3 2.9
Daily rate			11.09 losing.

III. EQUAL ALTITUDES OF A FIXED STAR.

The Observation.

In selecting stars for this observation, it is to be observed that the nearer the zenith the star passes, the less may the elapsed time be; and when the star passes exactly through the zenith, the two altitudes may be taken within a few minutes of each other. But with the ordinary sextants, altitudes near 90° cannot be taken with the artificial horizon, as the double altitude is then nearly 180°. The prismatic sextants, or still better, the prismatic circles of Pistor and Martin, are adapted for measuring angles of all magnitudes up to 180°, and are therefore especially suitable for this observation.

Set the sextant and wait for the coincidences of the two images of the star, as in the case of the sun's limb, noting the times by chronometer or watch.

The Computation.

Take the mean of the times before the meridian passage as the *A. M. Chronometer Time*, and the mean of those after the meridian passage as the *P. M. Chronometer Time*.

The mean of the A. M. and P. M. Chronometer Times is the *Chronometer Time of Star's Transit*. This time, if the chronometer is right, will agree with the true mean time of star's transit, which is to be computed as follows.

To the right ascension of the star apply the longitude of the place of observation (adding in west, subtracting in east); the result is the *Greenwich Sidereal Time of Star's Transit*, from which subtract the sidereal time at the *preceding* mean noon Greenwich (Nautical Almanac, page II. of the month); the remainder is the *Sidereal Interval* since mean noon. From Table IV. with the argument *Sidereal Interval*, take out the correction, which subtract from the sidereal interval; the remainder is the Greenwich Mean Time of the Star's Transit. The chronometer time will be more or less than this according as the chronometer is fast or slow.

If the chronometer is regulated to local time, apply the longitude to the Greenwich mean time of star's transit (subtracting in west, adding in east); the result is the *Local Mean Time of Star's Transit*, and the chronometer is fast or slow according as it shows more or less than this time.

Example 3.

July 15th, 1856, at the Cape of Good Hope, Lat. 33° 56′ S., Long. 1ʰ· 13ᵐ· 56′· E., observed equal altitudes of *Antares* as follows: —

EQUAL ALTITUDES.

Chronom. A. M.	2 Alt. Antares.	Chronom. P. M
h. m. s.	° ′	h. m. s.
5 32 10.5	125 30	9 34 20.3
5 32 35.0	40	9 33 56.0
5 32 59.3	50	9 33 32.0

A. M. Chro. T.	5 32 34.9		P. M. Chro. T.	9 33 56.1
P. M. Chro. T.	9 33 56.1			
	2)15 6 31.0			h. m. s.
			Antares R. A.	16 20 37.58
Chro. T. ✳ Transit	7 33 15.5		Longitude	1 13 56.00 E
Gr. T. ✳ Transit	7 31 22.1		Gr. Sid. T.	15 6 41.58
Chro. fast	1 53.4		July 15, Gr. Sid. T. Mean Noon	7 34 5.25
			Sid. Interval	7 32 36.33
			Correction, Table IV.	—1 14.15
			Gr. M. T. ✳ Transit	7 31 22.18

IV. TO CORRECT FOR SMALL INEQUALITIES IN THE ALTITUDES

Although the sextant readings are the same at the A. M. and P. M. observations, it may happen that neither the true nor even the apparent altitudes are the same. 1st. Supposing the sextant to remain unchanged, the atmospheric refraction may be different at the two observations in consequence of changes in the density and temperature of the air as shown by the barometer and thermometer. In this case, the apparent altitudes are equal, but the true altitudes are not so. 2d. The sextant may be affected by changes of temperature, particularly in day observations in the sun, so as to make the sextant readings the same for apparent altitudes slightly different. I do not think these changes in the sextant are to be eliminated by determining the index error at each observation, as has been supposed by some, since it is quite possible that the expansion and contraction of the various parts might leave the index correction unchanged while it affected the readings of the altitudes, or the reverse. The only course appears to be to guard the instrument as much as possible from changes of temperature, exposing it to the sun's rays only during the few minutes required for each observation.

But the correction for changes of refraction may be satisfactorily made as follows. Note the barometer and thermometer both A. M. and P. M.; take out the corresponding refractions for each observation from Tables III., III. A., and III. B., and find the difference of these refractions. Also take the difference between any two sextant readings and the difference between the two corresponding chronometer times. Then the correction of either noon or midnight will be found by the following proportion. The difference of the sextant readings is to the difference of the refractions as the difference of the chronometer times is to the required correction.

Apply this correction to the Chronometer Time of Noon or Midnight (obtained by the preceding rules) as follows: *add* it when the A. M. refraction is the greater; *subtract* it when the P. M. refraction is the greater. The result is the true Chronometer Time of Noon or Midnight.

EXAMPLE. — Suppose, in Example 1, we have in the morning, Barometer 30 inches, Thermometer 55°; in the afternoon, Barometer 29.5 inches, Thermometer 85°. The apparent altitude of sun's lower limb 33° 0′; the apparent altitude of sun's centre 33° 16′. We have

71

	A M.				P. M.	
Mean refraction	1 29			Mean refraction	1 29	
Barom. 30 in.	0			Barom. 29.5 in.	−1	
Therm. 55°	−1			Therm. 85°	−6	
True refraction	1 28			True refraction	1 22	

Then the difference of the sextant readings is 10′ (=600″) and the corresponding diff. of chronometer times is about 26′ ; whence

$$600'' : 6'' = 26^{s.\ s} : 0.26$$

The (approximate) Chronometer Time of Mean Noon was found to be   5 6 59.5 (h. m. s.)
Correction for change of refraction   +0.3
True Chronometer Time of Mean Noon   5 6 59.8

NOTE. — This correction may be found by the following rule, which we should have to resort to when but one altitude was taken at each observation. Add together the log. of the diff. of refractions (Tab. II.), log. cosine of the altitude, log. secant of the latitude, log. secant of the declination, log. cosecant of half elapsed time (or if the elapsed time is greater than 12ʰ, half its supplement to 24ʰ), and the constant log. 8.523; the sum is the log. (Table II.) of the required correction. Thus in the preceding example we have

Diff. refr.	6″	log.	0.778
Alt. ☉	33° 16′	log. cos.	9.922
Lat.	38° 59′	log. sec.	0.109
Dec.	15° 53′	log. sec.	0.017
El. T.	7ʰ 49ᵐ.	log. cosec.*	0.069
		const. log.	8.523
Correction	0ˢ.26	log.	9.418

DEGREE OF DEPENDENCE.

An error of 5′ in the latitude would not affect the corresponding part of the equation of equal altitudes by more than one hundredth of its amount in the most unfavorable case, and in general would have no sensible effect. It is one of the advantages of the equal altitude method, therefore, that it does not require an accurate knowledge of the latitude. It is also plain that errors in the longitude affecting the declination and its hourly difference produce but small proportionate effects upon the computed equation. The absolute error of the chronometer on Greenwich will be affected by the whole error in the longitude, but the *rate* will still be correct. Hence we conclude that by this method the chronometer may be accurately *rated* at a place whose latitude and longitude are both imperfectly known.

The chief source of error is in the observation itself. The most practised observers with the sextant cannot depend on the noted time of a *single* contact within 0ˢ.5, and hence the intervals between the successive chronometer times (which, if observations could be perfectly taken would be sensibly equal) may differ 2ˢ. But the greatest probable error of the chronometer time of sun's or star's transit, from the mean of six such observations on each side of the meridian, is found to be not more than 0ˢ.2, provided the rate of the chronometer between the observations is uniform.

Errors resulting from changes in the refraction may be almost wholly removed by computation as above.

* Enter BOWDITCH'S Table XXVII., column P. M., with the *whole* elapsed time and take out the corresponding cosecant.

72

EXPLANATION OF THE TABLES.

TABLE I. — *Logarithms of A and B, for computing the Equation of Equal Altitudes*, are calculated by the formulas

$$A = \frac{E}{1800 \sin \frac{1}{2} E}, \quad B = \frac{E}{1800 \tan \frac{1}{2} E},$$

where E = elapsed time in minutes, and E in the denominator is the elapsed time expressed in arc.

If we put

ϕ = latitude of the place of observation, + north, — south,
δ = declination of the sun,   + north, — south,
Δ = hourly change of declination,   + north, — south,
χ = correction to reduce the middle chronometer time to chronometer time of apparent noon, algebraically additive,
χ' = the same for midnight,

we have

$$\chi = - A \, \Delta \tan \phi + B \, \Delta \tan \delta$$
$$\chi' = A \, \Delta \tan \phi + B \, \Delta \tan \delta.$$

TABLE II. — *Logarithms of Numbers* to four decimal places. The first two figures of the number are found in the left-hand column, the third at the top, and the corresponding logarithm opposite and under these respectively. The proportional part for the fourth figure is found on the side in the same line with the logarithm taken out. The proper characteristic of the logarithm is to be supplied by the usual rule.

TABLE III. — *Mean Refraction*, reduced from BESSEL's Tables, to barometer 30 inches, and thermometer 50°.

TABLES III. A. and III. B. — *Corrections of the Mean Refraction for the Height of the Barometer and Thermometer*, also deduced from BESSEL's Tables. These are the same as Tables IV. A. and IV. B., given in the Appendix to the Nautical Almanac for 1855, where they are used for finding the corrections of the *Mean Reduced Refraction for Lunars*. It is for the purpose of having the same table for correcting both these mean refraction tables, that the argument in Tables III. A. and III. B. is the mean refraction instead of the apparent altitude.

TABLE IV. — *For converting Sidereal into Mean Solar Time*. This table gives the correction required to reduce a sidereal interval to its equivalent solar interval.

10   73

TABLE I. LOG. A. AND LOG. B.

For Computing the Equation of Equal Altitudes.

| For Noon, A —
For Midnight, A +} | ARGUMENT = ELAPSED TIME. | {For Noon or
Midnight, B + |

Elapsed Time.	0ʰ Log. A	0ʰ Log. B	1ʰ Log. A	1ʰ Log. B	2ʰ Log. A	2ʰ Log. B	3ʰ Log. A	3ʰ Log. B	4ʰ Log. A	4ʰ Log. B	5ʰ Log. A	5ʰ Log. B
m												
0	9.4059	9.4059	9.4072	9.4034	9.4109	9.3959	9.4172	9.3828	9.4260	9.3635	9.4374	9.3369
1	.4059	.4059	.4072	.4034	.4110	.3957	.4173	.3825	.4261	.3631	.4376	.3364
2	.4059	.4059	.4073	.4033	.4111	.3955	.4174	.3822	.4263	.3627	.4378	.3358
3	.4059	.4059	.4073	.4032	.4112	.3953	.4175	.3820	.4265	.3624	.4380	.3353
4	.4059	.4059	.4074	.4031	.4113	.3952	.4177	.3817	.4266	.3620	.4383	.3348
5	9.4059	9.4059	.4074	9.4030	.4113	9.3950	.4178	9.3814	.4268	9.3616	.4385	9.3343
6	.4060	.4059	.4074	.4029	.4114	.3948	.4179	.3811	.4270	.3612	.4387	.3337
7	.4060	.4059	.4075	.4028	.4115	.3946	.4181	.3809	.4272	.3608	.4389	.3332
8	.4060	.4059	.4075	.4027	.4116	.3944	.4182	.3806	.4273	.3604	.4391	.3327
9	.4060	.4059	.4076	.4026	.4117	.3943	.4183	.3803	.4275	.3600	.4393	.3321
10	9.4060	9.4059	.4076	9.4025	.4118	9.3941	.4184	9.3800	.4277	9.3596	.4396	9.3316
11	.4060	.4059	.4077	.4024	.4119	.3939	.4186	.3797	.4279	.3592	.4398	.3311
12	.4060	.4058	.4077	.4023	.4120	.3937	.4187	.3794	.4280	.3588	.4400	.3305
13	.4060	.4058	.4078	.4022	.4121	.3935	.4188	.3792	.4282	.3584	.4402	.3300
14	.4060	.4058	.4078	.4021	.4121	.3933	.4190	.3789	.4284	.3580	.4405	.3294
15	9.4060	9.4058	.4079	9.4020	.4122	9.3931	.4191	9.3786	.4286	9.3576	.4407	9.3289
16	.4060	.4058	.4079	.4019	.4123	.3929	.4193	.3783	.4288	.3572	.4409	.3283
17	.4060	.4057	.4080	.4018	.4124	.3927	.4194	.3780	.4289	.3568	.4411	.3278
18	.4061	.4057	.4080	.4017	.4125	.3925	.4195	.3777	.4291	.3564	.4414	.3272
19	.4061	.4057	.4081	.4016	.4126	.3923	.4197	.3774	.4293	.3559	.4416	.3266
20	9.4061	9.4057	.4081	9.4015	.4127	9.3921	.4198	9.3771	.4295	9.3555	.4418	9.3261
21	.4061	.4056	.4082	.4014	.4128	.3919	.4199	.3768	.4297	.3551	.4420	.3255
22	.4061	.4056	.4083	.4013	.4129	.3917	.4201	.3765	.4299	.3547	.4423	.3249
23	.4061	.4056	.4083	.4012	.4130	.3915	.4202	.3762	.4300	.3542	.4425	.3244
24	.4061	.4055	.4084	.4010	.4131	.3913	.4204	.3759	.4302	.3538	.4427	.3238
25	9.4062	9.4055	.4084	9.4009	.4132	9.3911	.4205	9.3756	.4304	9.3534	.4430	9.3232
26	.4062	.4055	.4085	.4008	.4133	.3909	.4207	.3752	.4306	.3530	.4432	.3226
27	.4062	.4054	.4086	.4007	.4134	.3907	.4208	.3749	.4308	.3525	.4434	.3220
28	.4062	.4054	.4086	.4006	.4135	.3905	.4209	.3746	.4310	.3521	.4437	.3214
29	.4062	.4054	.4087	.4004	.4136	.3903	.4211	.3743	.4312	.3516	.4439	.3208
30	9.4062	9.4053	.4087	9.4003	.4137	9.3900	.4212	9.3740	.4314	9.3512	.4441	9.3203
31	.4063	.4053	.4088	.4002	.4138	.3898	.4214	.3737	.4315	.3508	.4444	.3197
32	.4063	.4052	.4089	.4001	.4139	.3896	.4215	.3733	.4317	.3503	.4446	.3191
33	.4063	.4052	.4089	.3999	.4140	.3894	.4217	.3730	.4319	.3499	.4448	.3185
34	.4063	.4051	.4090	.3998	.4141	.3892	.4218	.3727	.4321	.3494	.4451	.3178
35	9.4064	9.4051	.4091	9.3997	.4142	9.3889	.4220	9.3723	.4323	9.3490	.4453	9.3172
36	.4064	.4050	.4091	.3995	.4144	.3887	.4221	.3720	.4325	.3485	.4456	.3166
37	.4064	.4050	.4092	.3994	.4145	.3885	.4223	.3717	.4327	.3480	.4458	.3160
38	.4064	.4049	.4093	.3993	.4146	.3882	.4224	.3713	.4329	.3476	.4460	.3154
39	.4065	.4049	.4093	.3991	.4147	.3880	.4226	.3710	.4331	.3471	.4463	.3148
40	9.4065	9.4048	.4094	9.3990	.4148	9.3878	.4227	9.3707	.4333	9.3467	.4465	9.3142
41	.4065	.4048	.4095	.3988	.4149	.3875	.4229	.3703	.4335	.3462	.4468	.3135
42	.4065	.4047	.4095	.3987	.4150	.3873	.4231	.3700	.4337	.3457	.4470	.3129
43	.4066	.4047	.4096	.3985	.4151	.3871	.4232	.3696	.4339	.3453	.4473	.3123
44	.4066	.4046	.4097	.3984	.4152	.3868	.4234	.3693	.4341	.3448	.4475	.3116
45	9.4066	9.4045	.4097	9.3982	.4154	9.3866	.4235	9.3690	.4343	9.3443	.4477	9.3110
46	.4067	.4045	.4098	.3981	.4155	.3863	.4237	.3686	.4345	.3438	.4480	.3103
47	.4067	.4044	.4099	3979	.4156	.3861	.4238	.3683	.4347	.3433	.4482	.3097
48	.4067	.4043	.4100	.3978	.4157	.3859	.4240	.3679	.4349	.3429	.4485	.3091
49	.4068	.4043	.4100	.3976	.4158	.3856	.4242	.3675	.4351	.3424	.4487	.3084
50	9.4068	9.4042	.4101	9.3975	.4159	9.3854	.4243	9.3672	.4353	9.3419	.4490	9.3078
51	.4068	.4041	.4102	.3973	.4161	.3851	.4245	.3668	.4355	.3414	.4492	.3071
52	.4069	.4041	.4103	.3972	.4162	.3849	.4246	.3665	.4357	.3409	.4494	.3064
53	.4069	.4040	.4103	.3970	.4163	.3846	.4248	.3661	.4359	.3404	.4497	.3058
54	.4069	.4039	.4104	.3969	.4164	.3843	.4250	.3657	.4361	.3399	.4500	.3051
55	9.4070	9.4038	.4105	9.3967	.4165	9.3841	.4251	9.3654	.4363	9.3394	.4503	9.3044
56	.4070	.4038	.4106	.3965	.4167	.3838	.4253	.3650	.4366	.3389	.4505	.3038
57	.4071	.4037	.4107	.3964	.4168	.3836	.4255	.3646	.4368	.3384	.4508	.3031
58	.4071	.4036	.4107	.3962	.4169	.3833	.4256	.3643	.4370	.3379	.4510	.3024
59	.4071	.4035	.4108	.3960	.4170	.3830	.4258	.3639	.4372	.3374	.4513	.3017
60	9.4072	9.4034	.4109	9.3959	.4172	9.3828	.4260	9.3635	.4374	9.3369	.4515	9.3010

74

TABLE I. LOG. A. AND LOG. B.

For Computing the Equation of Equal Altitudes.

| For Noon, A — / For Midnight, A +} | ARGUMENT = ELAPSED TIME. | {For Noon or / Midnight, B + |

Elapsed Time (m)	6ʰ Log. A.	6ʰ Log. B.	7ʰ Log. A.	7ʰ Log. B.	8ʰ Log. A.	8ʰ Log. B.	9ʰ Log. A.	9ʰ Log. B.	10ʰ Log. A.	10ʰ Log. B.	11ʰ Log. A.	11ʰ Log. B.
0	9.4515	9.3010	9.4685	9.2530	9.4884	9.1874	9.5115	9.0943	9.5379	8.9509	9.5680	8.6837
1	.4518	.3003	.4688	.2520	.4888	.1861	.5119	.0925	.5384	.9478	.5685	.6770
2	.4521	.2996	.4691	.2511	.4892	.1848	.5123	.0906	.5389	.9447	.5691	.6701
3	.4523	.2989	.4694	.2502	.4895	.1835	.5127	.0887	.5393	.9416	.5696	.6632
4	.4526	.2982	.4697	.2492	.4899	.1822	.5132	.0867	.5398	.9384	.5701	.6560
5	9.4528	9.2975	9.4701	9.2483	9.4902	9.1809	9.5136	9.0848	9.5403	8.9352	9.5707	8.6488
6	.4531	.2968	.4704	.2473	.4906	.1796	.5140	.0828	.5408	.9320	.5712	.6414
7	.4534	.2961	.4707	.2463	.4910	.1782	.5144	.0809	.5412	.9287	.5718	.6339
8	.4536	.2954	.4710	.2454	.4913	.1769	.5148	.0789	.5417	.9254	.5723	.6262
9	.4539	.2947	.4713	.2444	.4917	.1756	.5153	.0769	.5422	.9221	.5728	.6183
10	9.4542	9.2940	9.4716	9.2434	9.4921	9.1742	9.5157	9.0749	9.5427	8.9187	9.5734	8.6103
11	.4544	.2932	.4719	.2425	.4924	.1728	.5161	.0729	.5432	.9153	.5739	.6021
12	.4547	.2925	.4723	.2415	.4928	.1715	.5165	.0708	.5436	.9118	.5745	.5937
13	.4550	.2918	.4726	.2405	.4932	.1701	.5169	.0688	.5441	.9083	.5750	.5852
14	.4552	.2911	.4729	.2395	.4935	.1687	.5174	.0667	.5446	.9048	.5756	.5764
15	9.4555	9.2903	9.4732	9.2385	9.4939	9.1673	9.5178	9.0646	9.5451	8.9013	9.5761	8.5674
16	.4558	.2896	.4735	.2375	.4943	.1659	.5182	.0625	.5456	.8977	.5767	.5583
17	.4561	.2888	.4738	.2365	.4946	.1645	.5186	.0604	.5461	.8940	.5772	.5488
18	.4563	.2881	.4742	.2355	.4950	.1630	.5191	.0583	.5466	.8903	.5778	.5392
19	.4566	.2873	.4745	.2344	.4954	.1616	.5195	.0561	.5470	.8866	.5783	.5293
20	9.4569	9.2866	9.4748	9.2334	9.4958	9.1602	9.5199	9.0540	9.5475	8.8829	9.5789	8.5192
21	.4572	.2858	.4751	.2324	.4961	.1587	.5204	.0518	.5480	.8791	.5794	.5088
22	.4574	.2850	.4755	.2313	.4965	.1573	.5208	.0496	.5485	.8752	.5800	.4981
23	.4577	.2843	.4758	.2303	.4969	.1558	.5212	.0474	.5490	.8713	.5806	.4871
24	.4580	.2835	.4761	.2292	.4973	.1543	.5217	.0452	.5495	.8674	.5811	.4758
25	9.4583	9.2827	9.4764	9.2282	9.4977	9.1528	9.5221	9.0429	9.5500	8.8634	9.5817	8.4641
26	.4585	.2819	.4768	.2271	.4980	.1513	.5225	.0406	.5505	.8594	.5822	.4521
27	.4588	.2812	.4771	.2261	.4984	.1498	.5230	.0383	.5510	.8553	.5828	.4397
28	.4591	.2804	.4774	.2250	.4988	.1483	.5234	.0360	.5515	.8512	.5834	.4270
29	.4594	.2796	.4778	.2239	.4992	.1468	.5238	.0337	.5520	.8470	.5839	.4138
30	9.4597	9.2788	9.4781	9.2228	9.4996	9.1453	9.5243	9.0314	9.5525	8.8427	9.5845	8.4001
31	.4600	.2780	.4784	.2217	.5000	.1437	.5247	.0290	.5530	.8384	.5851	.3860
32	.4602	.2772	.4788	.2206	.5003	.1422	.5252	.0266	.5535	.8341	.5856	.3713
33	.4605	.2764	.4791	.2195	.5007	.1406	.5256	.0242	.5510	.8297	.5862	.3561
34	.4608	.2756	.4794	.2184	.5011	.1390	.5261	.0218	.5545	.8253	.5868	.3403
35	9.4611	9.2747	9.4799	9.2173	9.5015	9.1375	9.5265	9.0194	9.5550	8.8208	9.5874	8.3239
36	.4614	.2739	.4801	.2162	.5019	.1359	.5269	.0169	.5555	.8162	.5879	.3067
37	.4617	.2731	.4804	.2151	.5023	.1343	.5274	.0144	.5560	.8115	.5885	.2888
38	.4620	.2723	.4808	.2140	.5027	.1327	.5278	.0119	.5565	.8068	.5891	.2701
39	.4622	.2714	.4811	.2128	.5031	.1310	.5283	.0094	.5570	.8020	.5897	.2505
40	9.4625	9.2706	9.4815	9.2117	9.5035	9.1294	9.5287	9.0069	9.5576	8.7972	9.5902	8.2299
41	.4628	.2698	.4818	.2105	.5038	.1278	.5292	.0043	.5581	.7923	.5908	.2082
42	.4631	.2689	.4821	.2094	.5042	.1261	.5296	.0017	.5586	.7873	.5914	.1853
43	.4634	.2681	.4825	.2082	.5046	.1244	.5301	8.9991	.5591	.7823	.5920	.1611
44	.4637	.2672	.4828	.2070	.5050	.1228	.5305	.9965	.5596	.7772	.5926	.1354
45	9.4640	9.2664	9.4832	9.2059	9.5054	9.1211	9.5310	8.9938	9.5601	8.7720	9.5931	8.1080
46	.4643	.2655	.4835	.2047	.5058	.1194	.5315	.9911	.5606	.7668	.5937	.0786
47	.4646	.2646	.4839	.2035	.5062	.1177	.5319	.9884	.5612	.7614	.5943	.0470
48	.4649	.2638	.4842	.2023	.5066	.1159	.5324	.9857	.5617	.7560	.5949	.0128
49	.4652	.2629	.4846	.2011	.5070	.1142	.5328	.9830	.5622	.7505	.5955	7.9756
50	9.4655	9.2620	9.4849	9.1999	9.5074	9.1125	9.5333	8.9802	9.5627	8.7449	9.5961	7.9348
51	.4658	.2611	.4853	.1987	.5078	.1107	.5337	.9774	.5632	.7392	.5967	.8897
52	.4661	.2602	.4856	.1974	.5082	.1089	.5342	.9745	.5638	.7335	.5973	.8391
53	.4664	.2593	.4860	.1962	.5086	.1072	.5347	.9717	.5643	.7276	.5979	.7817
54	.4667	.2584	.4863	.1950	.5091	.1054	.5351	.9688	.5648	.7217	.5985	.7154
55	9.4670	9.2575	9.4867	9.1937	9.5095	9.1036	9.5356	8.9659	9.5654	8.7156	9.5991	7.6368
56	.4673	.2566	.4870	.1925	.5099	.1017	.5361	.9630	.5659	.7094	.5997	.5405
57	.4676	.2557	.4874	.1912	.5103	.0999	.5365	.9600	.5664	.7032	.6003	.4162
58	.4679	.2548	.4877	.1900	.5107	.0981	.5370	.9570	.5669	.6968	.6009	.2407
59	.4682	.2539	.4881	.1887	.5111	.0962	.5375	.9540	.5675	.6903	.6015	6.9591
60	9.4685	9.2530	9.4884	9.1874	9.5115	9.0943	9.5379	8.9509	9.5680	8.6837	9.6021	Inf.

TABLE I. LOG. A. AND LOG. B.

For Computing the Equation of Equal Altitudes.

For Noon, A — / For Midnight, A + }		ARGUMENT = ELAPSED TIME.									{ For Noon or Midnight, D —
Elapsed Time.	12ʰ.		13ʰ.		14ʰ.		15ʰ.		16ʰ.		17ʰ.	
m	Log. A.	Log. B.	Log. A.	Log. B	Log. A.	Log. B.	Log. A.	Log. B.	Log. A.	Log. B.	Log. A.	Log. B.
0	9.6021	$Inf.$	9.6406	8.7563	9.6841	9.0071	9.7333	9.3162	9.7895	9.4884	9.8539	9.6383
1	.6027	6.9603	.6412	.7641	.6848	.1014	.7342	.3194	.7905	.4911	.8550	.6407
2	.6033	7.2481	.6419	.7718	.6856	.1057	.7351	.3225	.7915	.4937	.8562	.6431
3	.6039	.4198	.6426	.7794	.6864	.1099	.7360	.3256	.7925	.4963	.8573	.6455
4	.6045	.5453	.6433	.7868	.6872	.1141	.7369	.3287	.7935	.4990	.8585	.6478
5	9.6051	7.6428	9.6440	8.7942	9.6879	9.1183	9.7378	9.3319	9.7945	9.5016	9.8597	9.6502
6	.6057	.7226	.6447	.8015	.6887	.1224	.7386	.3350	.7955	.5042	.8608	.6526
7	.6063	.7902	.6454	.8087	.6895	.1265	.7395	.3380	.7965	.5068	.8620	.6550
8	.6069	.8488	.6461	.8158	.6903	.1306	.7404	.3411	.7975	.5094	.8632	.6573
9	.6075	.9005	.6467	.8227	.6911	.1347	.7413	.3442	.7986	.5120	.8644	.6597
10	9.6082	7.9469	9.6474	8.8296	9.6919	9.1387	9.7422	9.3472	9.7996	9.5146	9.8655	9.6621
11	.6088	.9889	.6481	.8364	.6926	.1428	.7431	.3503	.8006	.5171	.8667	.6644
12	.6094	8.0273	.6488	.8432	.6934	.1468	.7440	.3533	.8016	.5197	.8679	.6668
13	.6100	.0627	.6495	.8498	.6942	.1507	.7449	.3563	.8027	.5223	.8691	.6691
14	.6106	.0955	.6502	.8564	.6950	.1547	.7458	.3593	.8037	.5248	.8703	.6715
15	9.6112	8.1260	9.6509	8.8628	9.6958	9.1586	9.7467	9.3623	9.8047	9.5274	9.8715	9.6738
16	.6119	.1547	.6516	.8692	.6966	.1625	.7476	.3653	.8058	.5300	.8727	.6762
17	.6125	.1816	.6523	.8756	.6974	.1664	.7485	.3683	.8068	.5325	.8739	.6785
18	.6131	.2071	.6530	.8818	.6982	.1703	.7494	.3713	.8078	.5351	.8751	.6809
19	.6137	.2312	.6538	.8880	.6990	.1741	.7503	.3742	.8089	.5376	.8763	.6832
20	9.6144	8.2541	9.6545	8.8941	9.6998	9.1779	9.7512	9.3772	9.8099	9.5401	9.8775	9.6856
21	.6150	.2759	.6552	.9002	.7006	.1817	.7522	.3801	.8110	.5427	.8787	.6879
22	.6156	.2967	.6559	.9062	.7014	.1855	.7531	.3831	.8120	.5452	.8799	.6903
23	.6163	.3166	.6566	.9121	.7022	.1893	.7540	.3860	.8131	.5477	.8812	.6926
24	.6169	.3357	.6573	.9180	.7030	.1930	.7549	.3889	.8141	.5502	.8824	.6949
25	9.6175	8.3540	9.6580	8.9238	9.7038	9.1967	9.7558	9.3918	9.8152	9.5528	9.8836	9.6973
26	.6182	.3717	.6588	.9295	.7047	.2004	.7568	.3947	.8162	.5553	.8848	.6996
27	.6188	.3887	.6595	.9352	.7055	.2041	.7577	.3976	.8173	.5578	.8861	.7019
28	.6194	.4051	.6602	.9408	.7063	.2078	.7586	.4005	.8184	.5603	.8873	.7043
29	.6201	.4210	.6609	.9464	.7071	.2114	.7595	.4033	.8194	.5628	.8885	.7066
30	9.6207	8.4363	9.6616	8.9519	9.7079	9.2150	9.7605	9.4062	9.8205	9.5653	9.8898	9.7089
31	.6214	.4512	.6624	.9573	.7088	.2186	.7614	.4090	.8216	.5677	.8910	.7112
32	.6220	.4657	.6631	.9627	.7096	.2222	.7624	.4119	.8227	.5702	.8923	.7136
33	.6226	.4796	.6638	.9681	.7104	.2258	.7633	.4147	.8237	.5727	.8935	.7159
34	.6233	.4932	.6645	.9734	.7112	.2293	.7642	.4175	.8248	.5752	.8948	.7182
35	9.6239	8.5064	9.6653	8.9787	9.7121	9.2329	9.7652	9.4204	9.8259	9.5777	9.8961	9.7205
36	.6246	.5192	.6660	.9839	.7129	.2364	.7661	.4232	.8270	.5801	.8973	.7228
37	.9252	.5318	.6667	.9891	.7137	.2399	.7671	.4260	.8281	.5826	.8986	.7251
38	.6259	.5440	.6675	.9942	.7146	.2434	.7680	.4288	.8292	.5850	.8999	.7275
39	.6265	.5559	.6682	.9993	.7154	.2468	.7690	.4316	.8303	.5875	.9011	.7298
40	9.6272	8.5675	9.6690	9.0043	9.7162	9.2503	9.7699	9.4343	9.8314	9.5900	9.9024	9.7321
41	.6279	.5788	.6697	.0093	.7171	.2537	.7709	.4371	.8325	.5924	.9037	.7344
42	.6285	.5899	.6704	.0142	.7179	.2571	.7718	.4399	.8336	.5948	.9050	.7367
43	.6292	.6008	.6712	.0191	.7187	.2605	.7728	.4426	.8347	.5973	.9063	.7390
44	.6298	.6114	.6719	.0240	.7196	.2639	.7738	.4454	.8358	.5997	.9075	.7413
45	9.6305	8.6218	9.6727	9.0288	9.7204	9.2673	9.7747	9.4481	9.8369	9.6022	9.9088	9.7436
46	.6311	.6320	.6734	.0336	.7213	.2706	.7757	.4509	.8380	.6046	.9101	.7459
47	.6318	.6419	.6742	.0384	.7221	.2740	.7767	.4536	.8391	.6070	.9114	.7482
48	.6325	.6517	.6749	.0431	.7230	.2773	.7776	.4563	.8402	.6094	.9127	.7505
49	.6331	.6613	.6757	.0478	.7238	.2806	.7786	.4590	.8414	.6119	.9140	.7529
50	9.6338	8.6707	9.6764	9.0524	9.7247	9.2839	9.7796	9.4617	9.8425	9.6143	9.9154	9.7552
51	.6345	.6799	.6772	.0570	.7256	.2872	.7806	.4644	.8436	.6167	.9167	.7575
52	.6351	.6890	.6779	.0616	.7264	.2905	.7815	.4671	.8447	.6191	.9180	.7598
53	.6358	.6979	.6787	.0662	.7273	.2937	.7825	.4698	.8459	.6215	.9193	.7621
54	.6365	.7067	.6795	.0707	.7281	.2970	.7835	.4725	.8470	.6239	.9206	.7644
55	9.6372	8.7153	9.6802	9.0752	9.7290	9.3002	9.7845	9.4752	9.8481	9.6263	9.9220	9.7667
56	.6378	.7237	.6810	.0796	.7299	.3034	.7855	.4778	.8493	.6287	.9233	.7690
57	.6385	.7321	.6818	.0840	.7307	.3066	.7865	.4805	.8504	.6311	.9246	.7713
58	.6392	.7402	.6825	.0884	.7316	.3098	.7875	.4831	.8516	.6335	.9260	.7736
59	.6399	.7483	.6833	.0928	.7324	.3130	.7885	.4858	.8527	.6359	.9273	.7759
60	9.6406	8.7563	9.6841	9.0971	9.7333	9.3162	9.7895	9.4884	9.8539	9.6383	9.9287	9.7782

TABLE I. LOG. A. AND LOG. B.

For Computing the Equation of Equal Altitudes.

For Noon, A — For Midnight, A+ }	ARGUMENT = ELAPSED TIME.									{ For Noon or Midnight, B —	
	18ʰ		19ʰ		20ʰ		21ʰ		22ʰ		2⅔ʰ
Elapsed Time. m	Log. A.	Log. B.	Log. A.	Log. B	Log. A.	Log. B.	Log. A.	Log. B.	Log. A.	Log. B.	Log. A.	Log. B.
0	9.9287	9.7782	0.0172	9.9167	0.1249	0.0625	0.2623	0.2279	0.4523	0.4372	0.7689	0.7652
1	.9300	.7804	.0188	.9190	.1269	.0650	.2649	.2309	.4562	.4414	.7765	.7729
2	.9314	.7827	.0204	.9213	.1290	.0676	.2676	.2339	.4601	.4455	.7842	.7807
3	.9327	.7850	.0221	.9237	.1310	.0701	.2702	.2370	.4640	.4497	.7920	.7886
4	.9341	.7873	.0237	.9260	.1330	.0727	.2729	.2401	.4680	.4540	.8000	.7967
5	9.9355	9.7896	0.0253	9.9284	0.1351	0.0753	0.2756	0.2431	0.4720	0.4582	0.8081	0.8049
6	.9368	.7919	.0270	.9307	.1371	.0779	.2783	.2462	.4761	.4625	.8163	.8133
7	.9382	.7942	.0286	.9331	.1392	.0805	.2810	.2493	.4801	.4668	.8247	.8218
8	.9396	.7965	.0303	.9355	.1412	.0830	.2838	.2524	.4842	.4711	.8333	.8305
9	.9410	.7988	.0319	.9378	.1433	.0856	.2865	.2556	.4884	.4755	.8420	.8393
10	9.9424	9.8011	0.0336	9.9402	0.1454	0.0882	0.2893	0.2587	0.4926	0.4799	0.8508	0.8483
11	.9437	.8034	.0353	.9426	.1475	.0909	.2921	.2619	.4968	.4844	.8599	.8574
12	.9451	.8057	.0370	.9449	.1496	.0935	.2949	.2650	.5010	.4889	.8691	.8667
13	.9465	.8080	.0386	.9473	.1517	.0961	.2977	.2682	.5053	.4934	.8786	.8763
14	.9479	.8103	.0403	.9497	.1538	.0987	.3005	.2714	.5097	.4980	.8882	.8860
15	9.9493	9.8126	0.0420	9.9520	0.1559	0.1013	0.3034	0.2746	0.5140	0.5026	0.8980	0.8959
16	.9508	.8149	.0437	.9544	.1581	.1040	.3063	.2778	.5184	.5072	.9080	.9060
17	.9522	.8172	.0454	.9568	.1602	.1066	.3091	.2811	.5229	.5118	.9183	.9164
18	.9536	.8195	.0472	.9592	.1623	.1093	.3120	.2843	.5274	.5165	.9288	.9270
19	.9550	.8218	.0489	.9616	.1645	.1119	.3150	.2876	.5319	.5213	.9396	.9378
20	9.9564	9.8241	0.0506	9.9640	0.1667	0.1146	0.3179	0.2909	0.5365	0.5261	0.9506	0.9489
21	.9579	.8264	.0523	.9664	.1689	.1173	.3208	.2942	.5411	.5309	.9618	.9603
22	.9593	.8287	.0541	.9687	.1711	.1200	.3238	.2975	.5458	.5358	.9734	.9719
23	.9607	.8310	.0558	.9711	.1733	.1226	.3268	.3008	.5505	.5407	.9853	.9839
24	.9622	.8333	.0576	.9735	.1755	.1253	.3298	.3041	.5553	.5457	.9975	.9961
25	9.9636	9.8356	0.0593	9.9760	0.1777	0.1280	0.3328	0.3075	0.5601	0.5507	1.0100	1.0087
26	.9651	.8379	.0611	.9784	.1799	.1308	.3359	.3109	.5649	.5557	.0228	.0216
27	.9665	.8402	.0628	.9808	.1821	.1335	.3389	.3143	.5698	.5608	.0361	.0350
28	.9680	.8425	.0646	.9832	.1844	.1362	.3420	.3177	.5748	.5660	.0497	.0487
29	.9695	.8448	.0664	.9856	.1867	.1389	.3451	.3211	.5798	.5712	.0638	.0628
30	9.9709	9.8471	0.0682	9.9880	0.1889	0.1417	0.3482	0.3245	0.5848	0.5764	1.0783	1.0774
31	.9724	.8494	.0700	.9904	.1912	.1444	.3514	.3280	.5899	.5817	.0934	.0925
32	.9739	.8517	.0718	.9929	.1935	.1472	.3545	.3315	.5951	.5871	.1089	.1081
33	.9754	.8540	.0736	.9953	.1958	.1499	.3577	.3350	.6003	.5925	.1250	.1242
34	.9769	.8563	.0754	.9977	.1981	.1527	.3609	.3385	.6056	.5979	.1416	.1409
35	9.9784	9.8586	0.0772	0.0002	0.2004	0.1555	0.3641	0.3420	0.6110	0.6034	1.1590	1.1583
36	.9798	.8609	.0790	.0026	.2028	.1582	.3674	.3456	.6164	.6090	.1770	.1764
37	.9813	.8632	.0809	.0051	.2051	.1610	.3706	.3491	.6218	.6147	.1958	.1952
38	.9829	.8655	.0827	.0075	.2075	.1638	.3739	.3527	.6273	.6204	.2154	.2149
39	.9844	.8678	.0845	.0100	.2098	.1667	.3772	.3563	.6329	.6261	.2359	.2354
40	9.9859	9.8701	0.0864	0.0124	0.2122	0.1695	0.3805	0.3599	0.6386	0.6319	1.2573	1.2569
41	.9874	.8724	.0883	.0149	.2146	.1723	.3839	.3636	.6443	.6378	.2799	.2795
42	.9889	.8748	.0901	.0173	.2170	.1751	.3873	.3673	.6501	.6438	.3037	.3033
43	.9904	.8771	.0920	.0198	.2194	.1780	.3907	.3710	.6560	.6498	.3288	.3285
44	.9920	.8794	.0939	.0223	.2218	.1808	.3941	.3747	.6619	.6559	.3554	.3552
45	9.9935	9.8817	0.0958	0.0248	0.2243	0.1837	0.3975	0.3784	0.6679	0.6621	1.3837	1.3835
46	.9951	.8840	.0976	.0272	.2267	.1866	.4010	.3822	.6740	.6684	.4140	.4138
47	.9966	.8863	.0995	.0297	.2292	.1895	.4045	.3859	.6802	.6747	.4465	.4463
48	.9982	.8887	.1015	.0322	.2316	.1924	.4080	.3897	.6865	.6811	.4815	.4814
49	.9998	.8910	.1034	.0347	.2341	.1953	.4115	.3936	.6928	.6876	.5196	.5195
50	0.0013	9.8933	0.1053	0.0372	0.2366	0.1982	0.4151	0.3974	0.6993	0.6942	1.5613	1.5612
51	.0029	.8956	.1072	.0397	.2391	.2011	.4187	.4013	.7058	.7008	.6074	.6073
52	.0044	.8980	.1092	.0422	.2416	.2040	.4223	.4052	.7124	.7076	.6588	.6587
53	.0060	.9003	.1111	.0447	.2442	.2070	.4260	.4091	.7191	.7144	.7171	.7171
54	.0076	.9026	.1131	.0473	.2467	.2099	.4297	.4130	.7259	.7214	.7844	.7843
55	0.0092	9.9050	0.1150	0.0498	0.2493	0.2129	0.4334	0.4170	0.7328	0.7284	1.8638	1.8638
56	.0108	.9073	.1170	.0523	.2518	.2159	.4371	.4210	.7398	.7355	.9610	.9610
57	.0124	.9096	.1190	.0548	.2544	.2189	.4408	.4250	.7469	.7428	2.0863	2.0863
58	.0140	.9120	.1209	.0574	.2570	.2219	.4446	.4291	.7541	.7501	.2627	.2627
59	.0156	.9143	.1229	.0599	.2596	.2249	.4485	.4331	.7615	.7576	2.5640	2.5640
60	0.0172	9.9167	0.1249	0.0625	0.2623	0.2279	0.4523	0.4372	0.7689	0.7652	Inf.	Inf.

TABLE II.

LOGARITHMS OF NUMBERS.

Natural Numbers	0	1	2	3	4	5	6	7	8	9	Proportional Parts.								
											1	2	3	4	5	6	7	8	9
10	0000	0043	0086	0128	0170	0212	0253	0294	0334	0374	4	8	12	17	21	25	29	33	37
11	0414	0453	0492	0531	0569	0607	0645	0682	0719	0755	4	8	11	15	19	23	26	30	34
12	0792	0828	0864	0899	0934	0969	1004	1038	1072	1106	3	7	10	14	17	21	24	28	31
13	1139	1173	1206	1239	1271	1303	1335	1367	1399	1430	3	6	10	13	16	19	23	26	29
14	1461	1492	1523	1553	1584	1614	1644	1673	1703	1732	3	6	9	12	15	18	21	24	27
15	1761	1790	1818	1847	1875	1903	1931	1959	1987	2014	3	6	8	11	14	17	20	22	25
16	2041	2068	2095	2122	2148	2175	2201	2227	2253	2279	3	5	8	11	13	16	18	21	24
17	2304	2330	2355	2380	2405	2430	2455	2480	2504	2529	2	5	7	10	12	15	17	20	22
18	2553	2577	2601	2625	2648	2672	2695	2718	2742	2765	2	5	7	9	12	14	16	19	21
19	2788	2810	2833	2856	2878	2900	2923	2945	2967	2989	2	4	7	9	11	13	16	18	20
20	3010	3032	3054	3075	3096	3118	3139	3160	3181	3201	2	4	6	8	11	13	15	17	19
21	3222	3243	3263	3284	3304	3324	3345	3365	3385	3404	2	4	6	8	10	12	14	16	18
22	3424	3444	3464	3483	3502	3522	3541	3560	3579	3598	2	4	6	8	10	12	14	15	17
23	3617	3636	3655	3674	3692	3711	3729	3747	3766	3784	2	4	6	7	9	11	13	15	17
24	3802	3820	3838	3856	3874	3892	3909	3927	3945	3962	2	4	5	7	9	11	12	14	16
25	3979	3997	4014	4031	4048	4065	4082	4099	4116	4133	2	3	5	7	9	10	12	14	15
26	4150	4166	4183	4200	4216	4232	4249	4265	4281	4298	2	3	5	7	8	10	11	13	15
27	4314	4330	4346	4362	4378	4393	4409	4425	4440	4456	2	3	5	6	8	9	11	13	14
28	4472	4487	4502	4518	4533	4548	4564	4579	4594	4609	2	3	5	6	8	9	11	12	14
29	4624	4639	4654	4669	4683	4698	4713	4728	4742	4757	1	3	4	6	7	9	10	12	13
30	4771	4786	4800	4814	4829	4843	4857	4871	4886	4900	1	3	4	6	7	9	10	11	13
31	4911	4928	4942	4955	4969	4983	4997	5011	5024	5038	1	3	4	6	7	8	10	11	12
32	5051	5065	5079	5092	5105	5119	5132	5145	5159	5172	1	3	4	5	7	8	9	11	12
33	5185	5198	5211	5224	5237	5250	5263	5276	5289	5302	1	3	4	5	6	8	9	10	12
34	5315	5328	5340	5353	5366	5378	5391	5403	5416	5428	1	3	4	5	6	8	9	10	11
35	5441	5453	5465	5478	5490	5502	5514	5527	5539	5551	1	2	4	5	6	7	9	10	11
36	5563	5575	5587	5599	5611	5623	5635	5647	5658	5670	1	2	4	5	6	7	8	10	11
37	5682	5694	5705	5717	5729	5740	5752	5763	5775	5786	1	2	3	5	6	7	8	9	10
38	5798	5809	5821	5832	5843	5855	5866	5877	5888	5899	1	2	3	5	6	7	8	9	10
39	5911	5922	5933	5944	5955	5966	5977	5988	5999	6010	1	2	3	4	5	7	8	9	10
40	6021	6031	6042	6053	6064	6075	6085	6096	6107	6117	1	2	3	4	5	6	8	9	10
41	6128	6138	6149	6160	6170	6180	6191	6201	6212	6222	1	2	3	4	5	6	7	8	9
42	6232	6243	6253	6263	6274	6284	6294	6304	6314	6325	1	2	3	4	5	6	7	8	9
43	6335	6345	6355	6365	6375	6385	6395	6405	6415	6425	1	2	3	4	5	6	7	8	9
44	6435	6444	6454	6464	6474	6484	6493	6503	6513	6522	1	2	3	4	5	6	7	8	9
45	6532	6542	6551	6561	6571	6580	6590	6599	6609	6618	1	2	3	4	5	6	7	8	9
46	6628	6637	6646	6656	6665	6675	6684	6693	6702	6712	1	2	3	4	5	6	7	7	8
47	6721	6730	6739	6749	6758	6767	6776	6785	6794	6803	1	2	3	4	5	5	6	7	8
48	6812	6821	6830	6839	6848	6857	6866	6875	6884	6893	1	2	3	4	4	5	6	7	8
49	6902	6911	6920	6928	6937	6946	6955	6964	6972	6981	1	2	3	4	4	5	6	7	8
50	6990	6998	7007	7016	7024	7033	7042	7050	7059	7067	1	2	3	3	4	5	6	7	8
51	7076	7084	7093	7101	7110	7118	7126	7135	7143	7152	1	2	3	3	4	5	6	7	8
52	7160	7168	7177	7185	7193	7202	7210	7218	7226	7235	1	2	2	3	4	5	6	7	7
53	7243	7251	7259	7267	7275	7284	7292	7300	7308	7316	1	2	2	3	4	5	6	6	7
54	7324	7332	7340	7348	7356	7364	7372	7380	7388	7396	1	2	2	3	4	5	6	6	7

TABLE II.

LOGARITHMS OF NUMBERS.

Natural Numbers	0	1	2	3	4	5	6	7	8	9	1	2	3	4	5	6	7	8	9
											Proportional Parts.								
55	7404	7412	7419	7427	7435	7443	7451	7459	7466	7474	1	2	2	3	4	5	5	6	7
56	7482	7490	7497	7505	7513	7520	7528	7536	7543	7551	1	2	2	3	4	5	5	6	7
57	7559	7566	7574	7582	7589	7597	7604	7612	7619	7627	1	2	2	3	4	5	5	6	7
58	7634	7642	7649	7657	7664	7672	7679	7686	7694	7701	1	1	2	3	4	4	5	6	7
59	7709	7716	7723	7731	7738	7745	7752	7760	7767	7774	1	1	2	3	4	4	5	6	7
60	7782	7789	7796	7803	7810	7818	7825	7832	7839	7846	1	1	2	3	4	4	5	6	6
61	7853	7860	7868	7875	7882	7889	7896	7903	7910	7917	1	1	2	3	4	4	5	6	6
62	7924	7931	7938	7945	7952	7959	7966	7973	7980	7987	1	1	2	3	3	4	5	6	6
63	7993	8000	8007	8014	8021	8028	8035	8041	8048	8055	1	1	2	3	3	4	5	5	6
64	8062	8069	8075	8082	8089	8096	8102	8109	8116	8122	1	1	2	3	3	4	5	5	6
65	8129	8136	8142	8149	8156	8162	8169	8176	8182	8189	1	1	2	3	3	4	5	5	6
66	8195	8202	8209	8215	8222	8228	8235	8241	8248	8254	1	1	2	3	3	4	5	5	6
67	8261	8267	8274	8280	8287	8293	8299	8306	8312	8319	1	1	2	3	3	4	5	5	6
68	8325	8331	8338	8344	8351	8357	8363	8370	8376	8382	1	1	2	3	3	4	4	5	6
69	8388	8395	8401	8407	8414	8420	8426	8432	8439	8445	1	1	2	2	3	4	4	5	6
70	8451	8457	8463	8470	8476	8482	8488	8494	8500	8506	1	1	2	2	3	4	4	5	6
71	8513	8519	8525	8531	8537	8543	8549	8555	8561	8567	1	1	2	2	3	4	4	5	5
72	8573	8579	8585	8591	8597	8603	8609	8615	8621	8627	1	1	2	2	3	4	4	5	5
73	8633	8639	8645	8651	8657	8663	8669	8675	8681	8686	1	1	2	2	3	4	4	5	5
74	8692	8698	8704	8710	8716	8722	8727	8733	8739	8745	1	1	2	2	3	4	4	5	5
75	8751	8756	8762	8768	8774	8779	8785	8791	8797	8802	1	1	2	2	3	3	4	5	5
76	8808	8814	8820	8825	8831	8837	8842	8848	8854	8859	1	1	2	2	3	3	4	5	5
77	8865	8871	8876	8882	8887	8893	8899	8904	8910	8915	1	1	2	2	3	3	4	4	5
78	8921	8927	8932	8938	8943	8949	8954	8960	8965	8971	1	1	2	2	3	3	4	4	5
79	8976	8982	8987	8993	8998	9004	9009	9015	9020	9025	1	1	2	2	3	3	4	4	5
80	9031	9036	9042	9047	9053	9058	9063	9069	9074	9079	1	1	2	2	3	3	4	4	5
81	9085	9090	9096	9101	9106	9112	9117	9122	9128	9133	1	1	2	2	3	3	4	4	5
82	9138	9143	9149	9154	9159	9165	9170	9175	9180	9186	1	1	2	2	3	3	4	4	5
83	9191	9196	9201	9206	9212	9217	9222	9227	9232	9238	1	1	2	2	3	3	4	4	5
84	9243	9248	9253	9258	9263	9269	9274	9279	9284	9289	1	1	2	2	3	3	4	4	5
85	9294	9299	9304	9309	9315	9320	9325	9330	9335	9340	1	1	2	2	3	3	4	4	5
86	9345	9350	9355	9360	9365	9370	9375	9380	9385	9390	1	1	2	2	3	3	4	4	5
87	9395	9400	9405	9410	9415	9420	9425	9430	9435	9440	0	1	1	2	2	3	3	4	4
88	9445	9450	9455	9460	9465	9469	9474	9479	9484	9489	0	1	1	2	2	3	3	4	4
89	9494	9499	9504	9509	9513	9518	9523	9528	9533	9538	0	1	1	2	2	3	3	4	4
90	9542	9547	9552	9557	9562	9566	9571	9576	9581	9586	0	1	1	2	2	3	3	4	4
91	9590	9595	9600	9605	9609	9614	9619	9624	9628	9633	0	1	1	2	2	3	3	4	4
92	9638	9643	9647	9652	9657	9661	9666	9671	9675	9680	0	1	1	2	2	3	3	4	4
93	9685	9689	9694	9699	9703	9708	9713	9717	9722	9727	0	1	1	2	2	3	3	4	4
94	9731	9736	9741	9745	9750	9754	9759	9763	9768	9773	0	1	1	2	2	3	3	4	4
95	9777	9782	9786	9791	9795	9800	9805	9809	9814	9818	0	1	1	2	2	3	3	4	4
96	9823	9827	9832	9836	9841	9845	9850	9854	9859	9863	0	1	1	2	2	3	3	4	4
97	9868	9872	9877	9881	9886	9890	9894	9899	9903	9908	0	1	1	2	2	3	3	4	4
98	9912	9917	9921	9926	9930	9934	9939	9943	9948	9952	0	1	1	2	2	3	3	4	4
99	9956	9961	9965	9969	9974	9978	9983	9987	9991	9996	0	1	1	2	2	3	3	3	4

TABLE III. MEAN REFRACTION.

Barometer 30 inches.   Fahrenheit's Thermometer 50°.

Apparent Altitude.	Mean Refraction.	Apparent Altitude.	Mean Refraction.	Apparent Altitude.	Mean Refraction.	Apparent Altitude.	Mean Refraction.	Apparent Altitude.	Mean Refraction.
° '	' "	° '	' "	° '	' "	° '	' "	° '	' "
		9 30	5 35.1	15 0	3 34.1	25 0	2 4.4	42 0	1 4.7
0 0	36 29.4	35	5 32.4	10	3 31.7	10	2 3.4	20	1 3.9
1 0	24 53.6	40	5 29.6	20	3 29.4	20	2 2.5	40	1 3.2
2 0	18 25.5	45	5 27.0	30	3 27.1	30	2 1.6	43 0	1 2.4
3 0	14 25.1	50	5 24.3	40	3 24.8	40	2 0.7	20	1 1.7
4 0	11 44.4	55	5 21.7	50	3 22.6	50	1 59.8	40	1 1.0
5 0	9 52.0	10 0	5 19.2	16 0	3 20.5	26 0	1 58.9	44 0	1 0.3
5	9 44.0	5	5 16.7	10	3 18.4	10	1 58.1	20	0 59.6
10	9 36.2	10	5 14.2	20	3 16.3	20	1 57.2	40	0 58.9
15	9 28.6	15	5 11.7	30	3 14.2	30	1 56.4	45 0	0 58.2
20	9 21.2	20	5 9.3	40	3 12.2	40	1 55.5	20	0 57.6
25	9 14.0	25	5 6.9	50	3 10.3	50	1 54.7	40	0 56.9
5 30	9 7.0	10 30	5 4.6	17 0	3 8.3	27 0	1 53.9	46 0	0 56.2
35	9 0.1	35	5 2.3	10	3 6.4	10	1 53.1	20	0 55.6
40	8 53.4	40	5 0.0	20	3 4.6	20	1 52.3	40	0 55.0
45	8 46.8	45	4 57.8	30	3 2.8	30	1 51.5	47 0	0 54.3
50	8 40.4	50	4 55.6	40	3 1.0	40	1 50.7	20	0 53.7
55	8 34.2	55	4 53.4	50	2 59.2	50	1 50.0	40	0 53.1
6 0	8 28.0	11 0	4 51.2	18 0	2 57.5	28 0	1 49.2	48 0	0 52.5
5	8 22.1	5	4 49.1	10	2 55.8	20	1 47.7	49 0	0 50.6
10	8 16.2	10	4 47.0	20	2 54.1	40	1 46.2	50 0	0 48.9
15	8 10.5	15	4 44.9	30	2 52.4	29 0	1 44.8	51 0	0 47.2
20	8 4.8	20	4 42.9	40	2 50.8	20	1 43.4	52 0	0 45.5
25	7 59.3	25	4 40.9	50	2 49.2	40	1 42.0	53 0	0 43.9
6 30	7 53.9	11 30	4 38.9	19 0	2 47.7	30 0	1 40.6	54 0	0 42.3
35	7 48.7	35	4 36.9	10	2 46.1	20	1 39.3	55 0	0 40.8
40	7 43.5	40	4 35.0	20	2 44.6	40	1 38.0	56 0	0 39.3
45	7 38.4	45	4 33.1	30	2 43.1	31 0	1 36.7	57 0	0 37.8
50	7 33.5	50	4 31.2	40	2 41.6	20	1 35.5	58 0	0 36.4
55	7 28.6	55	4 29.4	50	2 40.2	40	1 34.2	59 0	0 35.0
7 0	7 23.8	12 0	4 27.5	20 0	2 38.8	32 0	1 33.0	60 0	0 33.6
5	7 19.2	5	4 25.7	10	2 37.4	20	1 31.8	61 0	0 32.3
10	7 14.6	10	4 23.9	20	2 36.0	40	1 30.7	62 0	0 31.0
15	7 10.1	15	4 22.2	30	2 34.6	33 0	1 29.5	63 0	0 29.7
20	7 5.7	20	4 20.4	40	2 33.3	20	1 28.4	64 0	0 28.4
25	7 1.4	25	4 18.7	50	2 32.0	40	1 27.3	65 0	0 27.2
7 30	6 57.1	12 30	4 17.0	21 0	2 30.7	34 0	1 26.2	66 0	0 25.9
35	6 53.0	35	4 15.3	10	2 29.4	20	1 25.1	67 0	0 24.7
40	6 48.9	40	4 13.6	20	2 28.1	40	1 24.1	68 0	0 23.6
45	6 44.9	45	4 12.0	30	2 26.9	35 0	1 23.1	69 0	0 22.4
50	6 41.0	50	4 10.4	40	2 25.7	20	1 22.0	70 0	0 21.2
55	6 37.1	55	4 8.8	50	2 24.5	40	1 21.0	71 0	0 20.1
8 0	6 33.3	13 0	4 7.2	22 0	2 23.3	36 0	1 20.1	72 0	0 18.9
5	6 29.6	5	4 5.6	10	2 22.1	20	1 19.1	73 0	0 17.8
10	6 25.9	10	4 4.1	20	2 20.9	40	1 18.2	74 0	0 16.7
15	6 22.3	15	4 2.6	30	2 19.8	37 0	1 17.2	75 0	0 15.6
20	6 18.8	20	4 1.0	40	2 18.7	20	1 16.3	76 0	0 14.5
25	6 15.3	25	3 59.6	50	2 17.5	40	1 15.4	77 0	0 13.5
8 30	6 11.9	13 30	3 58.1	23 0	2 16.4	38 0	1 14.5	78 0	0 12.4
35	6 8.5	35	3 56.6	10	2 15.4	20	1 13.6	79 0	0 11.3
40	6 5.2	40	3 55.2	20	2 14.3	40	1 12.7	80 0	0 10.3
45	6 2.0	45	3 53.7	30	2 13.3	39 0	1 11.9	81 0	0 9.2
50	5 58.8	50	3 52.3	40	2 12.2	20	1 11.0	82 0	0 8.2
55	5 55.7	55	3 50.9	50	2 11.2	40	1 10.2	83 0	0 7.2
9 0	5 52.6	14 0	3 49.5	24 0	2 10.2	40 0	1 9.4	84 0	0 6.1
5	5 49.6	10	3 46.8	10	2 9.2	20	1 8.6	85 0	0 5.1
10	5 46.6	20	3 44.2	20	2 8.2	40	1 7.8	86 0	0 4.1
15	5 43.6	30	3 41.6	30	2 7.2	41 0	1 7.0	87 0	0 3.1
20	5 40.7	40	3 39.0	40	2 6.2	20	1 6.2	88 0	0 2.0
25	5 37.9	50	3 36.5	50	2 5.3	40	1 5.4	89 0	0 1.0
9 30	5 35.1	15 0	3 34.1	25 0	2 4.4	42 0	1 4.7	90 0	0 0.0

TABLE III. A.

Correction of the Mean Refraction for the Height of the Barometer.

Barometer. Subtract.	|				MEAN REFRACTION.						Barometer. Add.
	0′	1′	2′	3′	4′	5′	6′	7′	8′	9′	10′	
	0″ 30″	0″ 30″	0″ 30″	0″ 30″	0″ 30″	0″ 30″	0″ 30″	0″ 30″	0″ 30″	0″ 30″	0″	
27.50	0 2	5 7	10 12	15 17	20 23	25 28	30 33	35 38	40 43	45 48	51	
27.55	0 2	5 7	10 12	15 17	20 22	25 27	30 32	35 37	40 42	45 47	50	
27.60	0 2	5 7	10 12	14 17	19 22	24 27	29 31	34 36	39 41	44 46	49	
27.65	0 2	5 7	9 12	14 16	19 21	24 26	28 31	34 36	38 40	43 45	48	
27.70	0 2	5 7	9 11	14 16	18 21	23 25	28 30	32 35	37 39	42 44	47	
27.75	0 2	4 7	9 11	13 16	18 20	23 25	27 29	32 34	36 39	41 43	46	
27.80	0 2	4 7	9 11	13 15	18 20	22 24	27 29	31 33	35 38	40 42	45	
27.85	0 2	4 6	9 11	13 15	17 19	22 24	26 28	30 32	35 37	39 41	44	
27.90	0 2	4 6	9 10	13 15	17 19	21 23	25 27	30 32	34 36	38 40	43	
27.95	0 2	4 6	8 10	12 14	16 18	21 23	25 27	29 31	33 35	37 39	42	
28.00	0 2	4 6	8 10	12 14	16 18	20 22	24 26	28 30	32 34	36 38	41	
28.05	0 2	4 6	8 10	12 14	16 18	20 22	24 25	27 29	31 33	35 37	39	
28.10	0 2	4 6	8 9	11 13	15 17	19 21	23 25	27 29	31 33	34 36	38	
28.15	0 2	4 6	7 9	11 13	15 17	19 20	22 24	26 28	30 32	34 36	37	
28.20	0 2	4 5	7 9	11 13	14 16	18 20	22 24	25 27	29 31	33 35	36	
28.25	0 2	3 5	7 9	10 12	14 16	18 19	21 23	25 26	28 30	32 34	35	
28.30	0 2	3 5	7 8	10 12	14 15	17 19	21 22	24 26	27 29	31 33	34	
28.35	0 2	3 5	7 8	10 12	13 15	17 18	20 22	23 25	27 28	30 32	33	
28.40	0 2	3 5	6 8	10 11	13 14	16 18	19 21	23 24	26 27	29 31	32	
28.45	0 2	3 5	6 8	9 11	12 14	16 17	19 20	22 23	25 27	28 30	31	
28.50	0 1	3 4	6 7	9 10	12 14	15 17	18 20	21 23	24 26	27 29	30	31.50
28.55	0 1	3 4	6 7	9 10	12 13	15 16	17 19	20 22	23 25	26 28	29	31.45
28.60	0 1	3 4	6 7	8 10	11 13	14 15	17 18	20 21	23 24	25 27	28	31.40
28.65	0 1	3 4	5 7	8 9	11 12	14 15	16 18	19 20	22 23	25 26	27	31.35
28.70	0 1	3 4	5 6	8 9	10 12	13 14	16 17	18 20	21 22	24 25	26	31.30
28.75	0 1	2 4	5 6	7 9	10 11	13 14	15 16	18 19	20 21	23 24	25	31.25
28.80	0 1	2 4	5 6	7 8	10 11	12 14	14 16	17 18	19 21	22 23	24	31.20
28.85	0 1	2 3	5 6	7 8	9 10	12 13	14 15	16 17	19 20	21 22	23	31.15
28.90	0 1	2 3	4 5	7 8	9 10	11 12	13 14	16 17	18 19	20 21	22	31.10
28.95	0 1	2 3	4 5	6 7	8 9	11 12	13 14	15 16	17 18	19 20	21	31.05
29.00	0 1	2 3	4 5	6 7	8 9	10 11	12 13	14 15	16 17	18 19	20	31.00
29.05	0 1	2 3	4 5	6 7	8 9	10 11	11 12	13 14	15 16	17 18	19	30.95
29.10	0 1	2 3	4 4	5 6	7 8	9 10	11 12	13 14	15 15	16 17	18	30.90
29.15	0 1	2 3	3 4	5 6	7 8	9 9	10 11	12 13	14 15	15 16	17	30.85
29.20	0 1	2 2	3 4	5 6	6 7	8 9	10 10	11 12	13 14	15 15	16	30.80
29.25	0 1	1 2	3 4	4 5	6 7	8 8	9 10	11 11	12 13	14 14	15	30.75
29.30	0 1	1 2	3 3	4 5	6 6	7 8	8 9	10 11	11 12	13 13	14	30.70
29.35	0 1	1 2	3 3	4 5	5 6	7 7	8 9	9 10	10 11	12 13	13	30.65
29.40	0 1	1 2	2 3	4 4	5 5	6 7	7 8	8 9	10 10	11 12	12	30.60
29.45	0 1	1 2	2 3	3 4	4 5	6 6	7 7	8 8	9 9	10 11	11	30.55
29.50	0 0	1 1	2 2	3 3	4 5	5 6	6 7	7 8	8 9	9 10	10	30.50
29.55	0 0	1 1	2 2	3 3	4 4	5 5	6 6	7 7	8 8	8 9	9	30.45
29.60	0 0	1 1	2 2	2 3	3 4	4 4	5 5	6 6	7 7	8 8	8	30.40
29.65	0 0	1 1	1 2	2 2	3 3	4 4	4 5	5 5	6 6	6 7	7	30.35
29.70	0 0	1 1	1 1	2 2	2 3	3 3	4 4	4 5	5 5	5 6	6	30.30
29.75	0 0	0 1	1 1	1 2	2 2	3 3	3 3	4 4	4 4	5 5	5	30.25
29.80	0 0	0 1	1 1	1 1	2 2	2 2	3 3	3 3	3 4	4 4	4	30.20
29.85	0 0	0 0	1 1	1 1	1 1	2 2	2 2	2 2	2 3	3 3	3	30.15
29.90	0 0	0 0	0 0	1 1	1 1	1 1	1 1	1 2	2 2	2 2	2	30.10
29.95	0 0	0 0	0 0	0 0	0 0	1 1	1 1	1 1	1 1	1 1	1	30.05
30.00	0 0	0 0	0 0	0 0	00 0	0 0	0 0	0 0	0 0	0 0	0	30.00
Subtract.	0″ 30″	0″ 30″	0″ 30″	0″ 30″	0″ 30″	0″ 30″	0″ 30″	0″ 30″	0″ 30″	0″ 30″	0″	Add.
	0′	1′	2′	3′	4′	5′	6′	7′	8′	9′	10′	
Barometer.					MEAN REFRACTION.							Barometer.

TABLE III. B.

Correction of the Mean Refraction for the Height of the Thermometer.

MEAN REFRACTION.

Each minute column holds the two sub-values printed under the headers **0″** and **30″**. The **10** column holds a single value.

Thermom. Add.	0′ (0 · 30)	1′ (0 · 30)	2′ (0 · 30)	3′ (0 · 30)	4′ (0 · 30)	5′ (0 · 30)	6′ (0 · 30)	7′ (0 · 30)	8′ (0 · 30)	9′ (0 · 30)	10	Thermom. Add.
−10	0 4	8 12	16 20	24 28	33 37	41 46	50 55	60 65	70 75	80 85	90	−10
− 8	0 4	8 12	15 19	23 27	31 36	40 44	48 53	58 62	67 72	77 82	87	− 8
− 6	0 4	7 11	15 19	22 26	30 34	38 42	47 51	55 60	64 69	74 79	84	− 6
− 4	0 4	7 11	14 18	22 25	29 33	37 41	45 49	53 57	62 66	71 76	80	− 4
− 2	0 3	7 10	14 17	21 24	28 31	35 39	43 47	51 55	59 64	68 72	77	− 2
0	0 3	7 10	13 16	20 23	27 30	34 37	41 45	49 53	57 61	65 69	74	0
2	0 3	6 9	12 16	19 22	25 29	32 36	39 43	47 50	54 58	62 66	70	2
4	0 3	6 9	12 15	18 21	24 28	31 34	37 41	44 48	52 55	59 63	67	4
6	0 3	6 8	11 14	17 20	23 26	29 32	36 39	42 46	49 53	56 60	64	6
8	0 3	5 8	11 14	16 19	22 25	28 31	34 37	40 43	47 50	54 57	61	8
10	0 3	5 8	10 13	15 18	21 24	26 29	32 35	38 41	44 48	51 54	58	10
11	0 2	5 7	10 13	15 18	20 23	26 28	31 34	37 40	43 46	49 53	56	11
12	0 2	5 7	10 12	15 17	20 22	25 28	30 33	36 39	42 45	48 51	54	12
13	0 2	5 7	9 12	14 17	19 22	24 27	30 32	35 38	41 44	47 50	53	13
14	0 2	5 7	9 11	14 16	19 21	24 26	29 31	34 37	40 42	45 48	51	14
15	0 2	4 7	9 11	13 16	18 20	23 25	28 30	33 36	38 41	44 47	50	15
16	0 2	4 6	9 11	13 15	18 20	22 25	27 29	32 35	37 40	43 45	48	16
17	0 2	4 6	8 10	13 15	17 19	21 24	26 29	31 33	36 39	41 44	47	17
18	0 2	4 6	8 10	12 14	16 19	21 23	25 28	30 32	35 37	40 43	45	18
19	0 2	4 6	8 10	12 14	16 18	20 22	24 27	29 31	34 36	39 41	44	19
20	0 2	4 6	8 9	11 13	15 17	19 22	24 26	28 30	33 35	37 40	42	20
21	0 2	4 5	7 9	11 13	15 17	19 21	23 25	27 29	31 34	36 38	41	21
22	0 2	3 5	7 9	11 12	14 16	18 20	22 24	26 28	30 32	35 37	39	22
23	0 2	3 5	7 8	10 12	14 15	17 19	21 23	25 27	29 31	33 36	38	23
24	0 2	3 5	6 8	10 11	13 15	17 18	20 22	24 26	28 30	32 34	36	24
25	0 2	3 5	6 8	9 11	13 14	16 18	19 21	23 25	27 29	31 33	35	25
26	0 1	3 4	6 7	9 11	12 14	15 17	19 20	22 24	26 28	29 31	33	26
27	0 1	3 4	6 7	9 10	12 13	15 16	18 19	21 23	25 26	28 30	32	27
28	0 1	3 4	5 7	8 10	11 12	14 15	17 19	20 22	23 25	27 29	30	28
29	0 1	3 4	5 6	8 9	11 12	13 15	16 18	19 21	22 24	26 27	29	29
30	0 1	2 4	5 6	7 9	10 11	13 14	15 17	18 20	21 23	24 26	28	30
31	0 1	2 3	5 6	7 8	9 11	12 13	15 16	17 19	20 22	23 25	26	31
32	0 1	2 3	4 6	7 8	9 10	11 13	14 15	16 18	19 20	22 23	25	32
33	0 1	2 3	4 5	6 7	8 10	11 12	13 14	15 17	18 19	21 22	23	33
34	0 1	2 3	4 5	6 7	8 9	10 11	12 13	14 16	17 18	19 21	22	34
35	0 1	2 3	4 5	6 6	7 8	9 10	11 13	14 15	16 17	18 19	20	35
36	0 1	2 3	3 4	5 6	7 8	9 10	11 12	13 14	15 16	17 18	19	36
37	0 1	2 2	3 4	5 6	6 7	8 9	10 11	12 13	14 15	16 17	18	37
38	0 1	1 2	3 4	4 5	6 7	7 8	9 10	11 12	13 13	14 15	16	38
39	0 1	1 2	3 3	4 5	5 6	7 8	8 9	10 11	11 12	13 14	15	39
40	0 1	1 2	2 3	4 4	5 6	6 7	8 8	9 10	10 11	12 13	13	40
41	0 1	1 2	2 3	3 4	4 5	6 6	7 7	8 9	9 10	11 11	12	41
42	0 0	1 1	2 2	3 3	4 4	5 5	6 7	7 8	8 9	9 10	11	42
43	0 0	1 1	2 2	3 3	3 4	4 5	5 6	6 7	7 8	8 9	9	43
44	0 0	1 1	1 2	2 3	3 3	4 4	4 5	5 6	6 7	7 8	8	44
45	0 0	1 1	1 1	2 2	3 3	3 4	4 5	5 6	6 6	7		45
46	0 0	0 1	1 1	1 2	2 2	2 2	3 3	4 4	4 4	5 5	6	46
47	0 0	0 1	1 1	1 1	1 2	2 2	2 2	3 3	3 3	4 4	4	47
48	0 0	0 0	0 1	1 1	1 1	1 1	1 1	2 2	2 2	2 2	3	48
49	0 0	0 0	0 0	0 0	0 0	1 1	1 1	1 1	1 1	1 1	1	49
50	0 0	0 0	0 0	0 0	0 0	0 0	0 0	0 0	0 0	0 0	0	50
Add.	0 30	0 30	0 30	0 30	0 30	0 30	0 30	0 30	0 30	0 30	0	Add.
Thermom.	0′	1′	2′	3′	4′	5′	6′	7′	8′	9′	10′	Thermom.

MEAN REFRACTION.

TABLE III. B.

Correction of the Mean Refraction for the Height of the Thermometer.

Thermom. Subtract	0 (″0 ″30)	1′ (″0 ″30)	2′ (″0 ″30)	3′ (″0 ″30)	4′ (″0 ″30)	5′ (″0 ″30)	6′ (″0 ″30)	7′ (″0 ″30)	8′ (″0 ″30)	9′ (″0 ″30)	10′ (″0)	Thermom. Subtract
50	0 0	0 0	0 0	0 0	0 0	0 0	0 0	0 0	0 0	0 0	0	50
51	0 0	0 0	0 0	0 0	0 1	1 1	1 1	1 1	1 1	1 1	1	51
52	0 0	0 0	0 1	1 1	1 1	1 1	1 2	2 2	2 2	2 2	3	52
53	0 0	0 1	1 1	1 1	1 2	2 2	2 2	2 3	3 3	3 4	4	53
54	0 0	0 1	1 1	1 2	2 2	2 3	3 3	3 4	4 4	5 5	5	54
55	0 0	1 1	1 1	2 2	2 3	3 3	4 4	4 5	5 5	6 6	6	55
56	0 0	1 1	1 2	2 2	3 3	4 4	4 5	5 6	6 6	7 7	8	56
57	0 0	1 1	2 2	2 3	3 4	4 5	5 6	6 6	7 8	8 8	9	57
58	0 0	1 1	2 2	3 3	4 4	5 5	6 6	7 7	8 9	9 10	10	58
59	0 1	1 2	2 3	3 4	4 5	5 6	6 7	8 8	9 10	10 11	12	59
60	0 1	1 2	2 3	3 4	5 5	6 7	7 8	9 9	10 11	11 12	13	60
61	0 1	1 2	3 3	4 4	5 6	7 7	8 9	9 10	11 12	12 13	14	61
62	0 1	1 2	3 3	4 5	6 6	7 8	9 9	10 11	12 13	14 15	15	62
63	0 1	1 2	3 4	5 5	6 7	8 8	9 10	11 12	13 14	15 16	17	63
64	0 1	2 2	3 4	5 6	7 7	8 9	10 11	12 13	14 15	16 17	18	64
65	0 1	2 3	3 4	5 6	7 8	9 10	11 12	13 14	15 16	17 18	19	65
66	0 1	2 3	4 5	6 6	7 8	9 10	11 12	14 15	16 17	18 19	20	66
67	0 1	2 3	4 5	6 7	8 9	10 11	12 13	14 16	17 18	19 20	22	67
68	0 1	2 3	4 5	6 7	8 9	11 11	13 14	15 16	18 19	20 22	23	68
69	0 1	2 3	4 5	7 8	9 10	11 12	13 15	16 17	19 20	21 23	24	69
70	0 1	2 3	5 6	7 8	9 10	12 12	14 16	17 18	20 21	22 24	25	70
71	0 1	2 4	5 6	7 8	10 11	12 13	15 16	18 19	20 22	23 25	27	71
72	0 1	2 4	5 6	8 9	10 11	13 14	16 17	18 20	21 23	25 26	28	72
73	0 1	3 4	5 7	8 9	11 12	13 14	16 18	19 21	22 24	26 27	29	73
74	0 1	3 4	5 7	8 10	11 12	14 15	17 18	20 22	23 25	27 28	30	74
75	0 1	3 4	6 7	8 10	11 13	14 16	18 19	21 22	24 26	28 29	31	75
76	0 1	3 4	6 7	9 10	12 13	15 16	18 20	22 23	25 27	29 31	32	76
77	0 1	3 5	6 8	9 11	12 14	16 17	19 21	22 24	26 28	30 32	34	77
78	0 2	3 5	6 8	9 11	13 14	16 18	20 21	23 25	27 29	31 33	35	78
79	0 2	3 5	6 8	10 11	13 15	17 18	20 22	24 26	28 30	32 34	36	79
80	0 2	3 5	7 8	10 12	14 15	17 19	21 23	25 27	29 31	33 35	37	80
81	0 2	3 5	7 9	10 12	14 16	18 20	21 24	26 28	30 32	34 36	38	81
82	0 2	4 5	7 9	11 13	14 16	18 20	22 24	26 28	31 33	35 37	40	82
83	0 2	4 5	7 9	11 13	15 17	19 21	23 25	27 29	31 34	36 38	41	83
84	0 2	4 6	8 9	11 13	15 17	19 21	23 26	28 30	32 35	37 39	42	84
85	0 2	4 6	8 10	12 14	16 18	20 22	24 26	29 31	33 36	38 40	43	85
86	0 2	4 6	8 10	12 14	16 18	20 23	25 27	29 32	34 37	39 42	44	86
87	0 2	4 6	8 10	12 14	17 19	21 23	25 28	30 32	35 38	40 43	45	87
88	0 2	4 6	8 10	13 15	17 19	21 24	26 28	31 33	36 38	41 44	46	88
89	0 2	4 6	9 11	13 15	17 20	22 24	27 29	32 34	37 39	42 45	48	89
90	0 2	4 7	9 11	13 16	18 20	23 25	27 30	32 35	38 40	43 46	49	90
91	0 2	4 7	9 11	14 16	18 21	23 25	28 31	33 36	39 41	44 47	50	91
92	0 2	5 7	9 11	14 16	19 21	24 26	29 31	34 37	39 42	45 48	51	92
93	0 2	5 7	9 12	14 17	19 22	24 27	29 32	35 37	40 43	46 49	52	93
94	0 2	5 7	10 12	14 17	19 22	25 27	30 33	35 38	41 44	47 50	53	94
95	0 2	5 7	10 12	15 17	20 22	25 28	30 33	36 39	42 45	48 51	54	95
96	0 2	5 7	10 12	15 18	20 23	26 28	31 34	37 40	43 46	49 52	55	96
97	0 3	5 8	10 13	15 18	21 23	26 29	32 35	38 41	44 47	50 53	56	97
98	0 3	5 8	10 13	16 18	21 24	27 29	32 35	38 41	44 48	51 54	58	98
99	0 3	5 8	11 13	16 19	21 24	27 30	33 36	39 42	45 49	52 55	59	99
100	0 3	5 8	11 13	16 19	22 25	28 31	34 37	40 43	46 50	53 56	60	100
Subtract (″0 ″30)	0 30	0 30	0 30	0 30	0 30	0 30	0 30	0 30	0 30	0 30	0	Subtract
Thermom.	0	1′	2′	3′	4′	5′	6′	7′	8′	9′	10′	Thermom.

MEAN REFRACTION.

TABLE IV. SIDEREAL INTO MEAN SOLAR TIME.

Sidereal m.	0 h.	1 h.	2 h.	3 h.	4 h.	5 h.	6 h.	7 h.	Sec.	For Seconds
0	0 00.000	0 09.830	0 19.659	0 29.489	0 30.318	0 49.148	0 58.977	1 08.807		
1	0 00.164	0 09.993	0 19.823	0 29.653	0 39.482	0 49.312	0 59.141	1 08.971	1	0.003
2	0 00.328	0 10.157	0 19.987	0 29.816	0 39.646	0 49.475	0 59.305	1 09.135	2	.005
3	0 00.491	0 10.321	0 20.151	0 29.980	0 39.810	0 49.639	0 59.469	1 09.298	3	.008
4	0 00.655	0 10.485	0 20.314	0 30.144	0 39.974	0 49.803	0 59.633	1 09.462	4	.011
5	0 00.819	0 10.649	0 20.478	0 30.308	0 40.137	0 49.967	0 59.796	1 09.626	5	.014
6	0 00.983	0 10.813	0 20.642	0 30.472	0 40.301	0 50.131	0 59.960	1 09.790	6	.016
7	0 01.147	0 10.976	0 20.806	0 30.635	0 40.465	0 50.295	1 00.124	1 09.954	7	.019
8	0 01.311	0 11.140	0 20.970	0 30.799	0 40.629	0 50.458	1 00.288	1 10.118	8	.022
9	0 01.474	0 11.304	0 21.134	0 30.963	0 40.793	0 50.622	1 00.452	1 10.281	9	.025
10	0 01.638	0 11.468	0 21.297	0 31.127	0 40.956	0 50.786	1 00.616	1 10.445	10	.027
11	0 01.802	0 11.632	0 21.461	0 31.291	0 41.120	0 50.950	1 00.773	1 10.609	11	.030
12	0 01.966	0 11.795	0 21.625	0 31.455	0 41.284	0 51.114	1 00.943	1 10.773	12	.033
13	0 02.130	0 11.959	0 21.789	0 31.618	0 41.448	0 51.278	1 01.107	1 10.937	13	.025
14	0 02.294	0 12.123	0 21.953	0 31.782	0 41.612	0 51.441	1 01.271	1 11.100	14	.038
15	0 02.457	0 12.287	0 22.117	0 31.946	0 41.776	0 51.605	1 01.435	1 11.264	15	.041
16	0 02.621	0 12.451	0 22.280	0 32.110	0 41.939	0 51.769	1 01.599	1 11.428	16	.044
17	0 02.785	0 12.615	0 22.444	0 32.274	0 42.103	0 51.933	1 01.762	1 11.592	17	.046
18	0 02.949	0 12.778	0 22.608	0 32.438	0 42.267	0 52.097	1 01.926	1 11.756	18	.049
19	0 03.113	0 12.942	0 22.772	0 32.601	0 42.431	0 52.261	1 02.090	1 11.920	19	.052
20	0 03.277	0 13.106	0 22.936	0 32.765	0 42.595	0 52.424	1 02.254	1 12.083	20	.055
21	0 03.440	0 13.270	0 23.099	0 32.929	0 42.759	0 52.588	1 02.418	1 12.247	21	.057
22	0 03.604	0 13.434	0 23.263	0 33.093	0 42.922	0 52.752	1 02.582	1 12.411	22	.060
23	0 03.768	0 13.598	0 23.427	0 33.257	0 43.086	0 52.916	1 02.745	1 12.575	23	.063
24	0 03.932	0 13.761	0 23.591	0 33.420	0 43.250	0 53.080	1 02.909	1 12.739	24	.066
25	0 04.096	0 13.925	0 23.755	0 33.584	0 43.414	0 53.243	1 03.073	1 12.903	25	.068
26	0 04.259	0 14.089	0 23.919	0 33.748	0 43.578	0 53.407	1 03.237	1 13.066	26	.071
27	0 04.423	0 14.253	0 24.082	0 33.912	0 43.742	0 53.571	1 03.401	1 13.230	27	.074
28	0 04.587	0 14.417	0 24.246	0 34.076	0 43.905	0 53.735	1 03.564	1 13.394	28	.076
29	0 04.751	0 14.581	0 24.410	0 34.240	0 44.069	0 53.899	1 03.728	1 13.558	29	.079
30	0 04.915	0 14.744	0 24.574	0 34.403	0 44.233	0 54.063	1 03.892	1 13.722	30	.082
31	0 05.079	0 14.908	0 24.738	0 34.567	0 44.397	0 54.226	1 04.056	1 13.886	31	.085
32	0 05.242	0 15.072	0 24.902	0 34.731	0 44.561	0 54.390	1 04.220	1 14.049	32	.087
33	0 05.406	0 15.236	0 25.065	0 34.895	0 44.724	0 54.554	1 04.384	1 14.213	33	.090
34	0 05.570	0 15.400	0 25.229	0 35.059	0 44.888	0 54.718	1 04.547	1 14.377	34	.093
35	0 05.734	0 15.563	0 25.393	0 35.223	0 45.052	0 54.882	1 04.711	1 14.541	35	.096
36	0 05.898	0 15.727	0 25.557	0 35.386	0 45.216	0 55.046	1 04.875	1 14.705	36	.098
37	0 06.062	0 15.891	0 25.721	0 35.550	0 45.380	0 55.209	1 05.039	1 14.868	37	.101
38	0 06.225	0 16.055	0 25.885	0 35.714	0 45.544	0 55.373	1 05.203	1 15.032	38	.104
39	0 06.389	0 16.219	0 26.048	0 35.878	0 45.707	0 55.537	1 05.367	1 15.196	39	.106
40	0 06.553	0 16.383	0 26.212	0 36.042	0 45.871	0 55.701	1 05.530	1 15.360	40	.109
41	0 06.717	0 16.546	0 26.376	0 36.206	0 46.035	0 55.865	1 05.694	1 15.524	41	.112
42	0 06.881	0 16.710	0 26.540	0 36.369	0 46.199	0 56.028	1 05.858	1 15.688	42	.115
43	0 07.045	0 16.874	0 26.704	0 36.533	0 46.363	0 56.192	1 06.022	1 15.851	43	.117
44	0 07.208	0 17.038	0 26.867	0 36.697	0 46.527	0 56.356	1 06.186	1 16.015	44	.120
45	0 07.372	0 17.202	0 27.031	0 36.861	0 46.690	0 56.520	1 06.350	1 16.179	45	.123
46	0 07.536	0 17.366	0 27.195	0 37.025	0 46.854	0 56.684	1 06.513	1 16.343	46	.126
47	0 07.700	0 17.529	0 27.359	0 37.188	0 47.018	0 56.848	1 06.677	1 16.507	47	.129
48	0 07.864	0 17.693	0 27.523	0 37.352	0 47.182	0 57.011	1 06.841	1 16.671	48	.131
49	0 08.027	0 17.857	0 27.687	0 37.516	0 47.346	0 57.175	1 07.005	1 16.834	49	.134
50	0 08.191	0 18.021	0 27.850	0 37.680	0 47.510	0 57.339	1 07.169	1 16.998	50	.137
51	0 08.355	0 18.185	0 28.014	0 37.844	0 47.673	0 57.503	1 07.332	1 17.162	51	.139
52	0 08.519	0 18.349	0 28.178	0 38.008	0 47.837	0 57.667	1 07.496	1 17.326	52	.142
53	0 08.683	0 18.512	0 28.342	0 38.171	0 48.001	0 57.831	1 07.660	1 17.490	53	.145
54	0 08.847	0 18.676	0 28.506	0 38.335	0 48.165	0 57.994	1 07.824	1 17.654	54	.147
55	0 09.010	0 18.840	0 28.670	0 38.499	0 48.329	0 58.158	1 07.988	1 17.817	55	.150
56	0 09.174	0 19.004	0 28.833	0 38.663	0 48.492	0 58.322	1 08.152	1 17.981	56	.153
57	0 09.338	0 19.168	0 28.997	0 38.827	0 48.656	0 58.486	1 08.315	1 18.145	57	.156
58	0 09.502	0 19.331	0 29.161	0 38.991	0 48.820	0 58.650	1 08.479	1 18.309	58	.158
59	0 09.666	0 19.495	0 29.325	0 39.154	0 48.984	0 58.814	1 08.643	1 18.473	59	.161

TABLE IV. SIDEREAL INTO MEAN SOLAR TIME.

Sidereal m.	8 h	9 h	10 h	11 h	12 h	13 h	14 h	15 h	For Seconds.
m.	m. s.	m. s.	m. s.	m. s.	m. s.	m. s.	m. s.	m. s.	s. s.
0	1 18.636	1 28.466	1 38.296	1 48.125	1 57.955	2 07.784	2 17.614	2 27.443	
1	1 18.800	1 28.630	1 38.459	1 48.289	1 58.119	2 07.948	2 17.778	2 27.607	1 0.003
2	1 18.964	1 28.794	1 38.623	1 48.453	1 58.282	2 08.112	2 17.941	2 27.771	2 .005
3	1 19.128	1 28.958	1 38.787	1 48.617	1 58.446	2 08.276	2 18.105	2 27.935	3 .008
4	1 19.292	1 29.121	1 38.951	1 48.780	1 58.610	2 08.440	2 18.269	2 28.099	4 .011
5	1 19.456	1 29.285	1 39.115	1 48.944	1 58.774	2 08.603	2 18.433	2 28.263	5 .014
6	1 19.61	1 29.449	1 39.279	1 49.108	1 58.938	2 08.767	2 18.597	2 28.426	6 .016
7	1 19.783	1 29.613	1 39.442	1 49.272	1 59.101	2 08.931	2 18.761	2 28.590	7 .019
8	1 19.947	1 29.777	1 39.606	1 49.436	1 59.265	2 09.095	2 18.924	2 28.754	8 .022
9	1 20.111	1 29.940	1 39.778	1 49.600	1 59.429	2 09.259	2 19.088	2 28.918	9 .025
10	1 20.275	1 30.104	1 39.934	1 49.763	1 59.593	2 09.423	2 19.252	2 29.082	10 .027
11	1 20.439	1 30.268	1 40.098	1 49.927	1 59.757	2 09.586	2 19.416	2 29.245	11 .030
12	1 20.602	1 30.432	1 40.261	1 50.091	1 59.921	2 09.750	2 19.580	2 29.409	12 .033
13	1 20.766	1 30.596	1 40.425	1 50.255	2 00.084	2 09.914	2 19.744	2 29.573	13 .035
14	1 20.930	1 30.760	1 40.589	1 50.419	2 00.248	2 10.078	2 19.907	2 29.737	14 .038
15	1 21.094	1 30.923	1 40.753	1 50.583	2 00.412	2 10.242	2 20.071	2 29.901	15 .041
16	1 21.258	1 31.087	1 40.917	1 50.746	2 00.576	2 10.405	2 20.235	2 30.065	16 .044
17	1 21.422	1 31.251	1 41.081	1 50.910	2 00.740	2 10.569	2 20.399	2 30.228	17 .046
18	1 21.585	1 31.415	1 41.244	1 51.074	2 00.904	2 10.733	2 20.563	2 30.392	18 .049
19	1 21.749	1 31.579	1 41.408	1 51.238	2 01.067	2 10.897	2 20.727	2 30.556	19 .052
20	1 21.913	1 31.743	1 41.572	1 51.402	2 01.231	2 11.061	2 20.890	2 30.720	20 .055
21	1 22.077	1 31.906	1 41.736	1 51.565	2 01.395	2 11.225	2 21.054	2 30.884	21 .057
22	1 22.241	1 32.070	1 41.900	1 51.729	2 01.559	2 11.388	2 21.218	2 31.048	22 .060
23	1 22.404	1 32.234	1 42.064	1 51.893	2 01.723	2 11.552	2 21.382	2 31.211	23 .063
24	1 22.568	1 32.398	1 42.227	1 52.057	2 01.897	2 11.716	2 21.546	2 31.375	24 .066
25	1 22.732	1 32.562	1 42.391	1 52.221	2 02.050	2 11.880	2 21.709	2 31.539	25 .068
26	1 22.896	1 32.726	1 42.555	1 52.385	2 02.214	2 12.044	2 21.873	2 31.703	26 .071
27	1 23.060	1 32.889	1 42.719	1 52.548	2 02.378	2 12.208	2 22.037	2 31.867	27 .074
28	1 23.224	1 33.053	1 42.883	1 52.712	2 02.542	2 12.371	2 22.201	2 32.031	28 .076
29	1 23.387	1 33.217	1 43.047	1 52.876	2 02.706	2 12.535	2 22.365	2 32.194	29 .079
30	1 23.551	1 33.381	1 43.210	1 53.040	2 02.869	2 12.699	2 22.529	2 32.358	30 .082
31	1 23.715	1 33.545	1 43.374	1 53.204	2 03.033	2 12.863	2 22.692	2 32.522	31 .085
32	1 23.879	1 33.708	1 43.538	1 53.368	2 03.197	2 13.027	2 22.856	2 32.686	32 .087
33	1 24.043	1 33.872	1 43.702	1 53.531	2 03.361	2 13.191	2 23.020	2 32.850	33 .090
34	1 24.207	1 34.036	1 43.866	1 53.695	2 03.525	2 13.354	2 23.184	2 33.013	34 .093
35	1 24.370	1 34.200	1 44.029	1 53.859	2 03.699	2 13.518	2 23.348	2 33.177	35 .096
36	1 24.534	1 34.364	1 44.193	1 54.023	2 03.852	2 13.682	2 23.512	2 33.341	36 .098
37	1 24.698	1 34.528	1 44.357	1 54.187	2 04.016	2 13.846	2 23.675	2 33.505	37 .101
38	1 24.862	1 34.691	1 44.521	1 54.351	2 04.180	2 14.010	2 23.839	2 33.669	38 .104
39	1 25.026	1 34.855	1 44.685	1 54.514	2 04.344	2 14.173	2 24.003	2 33.833	39 .106
40	1 25.190	1 35.019	1 44.849	1 54.678	2 04.508	2 14.337	2 24.167	2 33.996	40 .109
41	1 25.353	1 35.183	1 45.012	1 54.842	2 04.672	2 14.501	2 24.331	2 34.160	41 .112
42	1 25.517	1 35.347	1 45.176	1 55.006	2 04.835	2 14.665	2 24.495	2 34.324	42 .115
43	1 25.681	1 35.511	1 45.340	1 55.170	2 04.999	2 14.829	2 24.658	2 34.488	43 .117
44	1 25.845	1 35.674	1 45.504	1 55.333	2 05.163	2 14.993	2 24.822	2 34.652	44 .120
45	1 26.009	1 35.839	1 45.668	1 55.497	2 05.327	2 15.156	2 24.986	2 34.816	45 .123
46	1 26.172	1 36.002	1 45.832	1 55.661	2 05.491	2 15.320	2 25.150	2 34.979	46 .125
47	1 26.336	1 36.166	1 45.995	1 55.655	2 05.655	2 15.484	2 25.314	2 35.143	47 .128
48	1 26.500	1 36.330	1 46.159	1 55.989	2 05.818	2 15.648	2 25.477	2 35.307	48 .131
49	1 26.664	1 36.493	1 46.323	1 56.153	2 05.982	2 15.812	2 25.641	2 35.471	49 .134
50	1 26.828	1 36.657	1 46.487	1 56.316	2 06.146	2 15.976	2 25.805	2 35.635	50 .137
51	1 26.992	1 36.821	1 46.651	1 56.480	2 06.310	2 16.139	2 25.969	2 35.798	51 .139
52	1 27.155	1 36.985	1 46.815	1 56.644	2 06.474	2 16.303	2 26.133	2 35.962	52 .142
53	1 27.319	1 37.149	1 46.978	1 56.808	2 06.637	2 16.467	2 26.297	2 36.126	53 .145
54	1 27.483	1 37.313	1 47.142	1 56.972	2 06.801	2 16.631	2 26.460	2 36.290	54 .147
55	1 27.647	1 37.476	1 47.306	1 57.136	2 06.965	2 16.795	2 26.624	2 36.454	55 .150
56	1 27.811	1 37.640	1 47.470	1 57.299	2 07.129	2 16.959	2 26.788	2 36.618	56 .153
57	1 27.975	1 37.804	1 47.634	1 57.463	2 07.293	2 17.122	2 26.952	2 36.781	57 .156
58	1 28.138	1 37.968	1 47.797	1 57.627	2 07.457	2 17.286	2 27.116	2 36.945	58 .158
59	1 28.302	1 38.132	1 47.961	1 57.791	2 07.620	2 17.450	2 27.280	2 37.109	59 .161

TABLE IV. SIDEREAL INTO MEAN SOLAR TIME.

Sidereal m	16 h	17 h	18 h	19 h	20 h	21 h	22 h	23 h	For Seconds
(m.) s.	m. s.	m. s.	m. s.	m. s.	m. s.	m. s.	m. s.	m. s.	s. / ."
0	2 37.273	2 47.102	2 56.932	3 06.762	3 16.591	3 26.421	3 36.250	3 46.080	
1	2 37.437	2 47.266	2 57.096	3 06.925	3 16.755	3 26.585	3 36.414	3 46.244	1 0.003
2	2 37.601	2 47.430	2 57.260	3 07.089	3 16.919	3 26.748	3 36.578	3 46.407	2 .005
3	2 37.764	2 47.594	2 57.424	3 07.253	3 17.083	3 26.912	3 36.742	3 46.571	3 .008
4	2 37.928	2 47.758	2 57.587	3 07.417	3 17.246	3 27.076	3 36.906	3 46.735	4 .011
5	2 38.092	2 47.922	2 57.751	3 07.581	3 17.410	3 27.240	3 37.069	3 46.899	5 .014
6	2 38.256	2 48.085	2 57.915	3 07.745	3 17.574	3 27.404	3 37.233	3 47.063	6 .016
7	2 38.420	2 48.249	2 58.079	3 07.908	3 17.738	3 27.568	3 37.397	3 47.227	7 .019
8	2 38.584	2 48.413	2 58.243	3 08.072	3 17.902	3 27.731	3 37.561	3 47.390	8 .022
9	2 38.747	2 48.577	2 58.406	3 08.236	3 18.066	3 27.895	3 37.725	3 47.554	9 .025
10	2 38.911	2 48.741	2 58.570	3 08.400	3 18.229	3 28.059	3 37.889	3 47.718	10 .027
11	2 39.075	2 48.905	2 58.734	3 08.564	3 18.393	3 28.223	3 38.052	3 47.882	11 .030
12	2 39.239	2 49.068	2 58.898	3 08.728	3 18.557	3 28.387	3 38.216	3 48.046	12 .033
13	2 39.403	2 49.232	2 59.062	3 08.891	3 18.721	3 28.550	3 38.380	3 48.210	13 .035
14	2 39.566	2 49.396	2 59.226	3 09.055	3 18.885	3 28.714	3 38.544	3 48.373	14 .038
15	2 39.730	2 49.560	2 59.389	3 09.219	3 19.049	3 28.878	3 38.708	3 48.537	15 .041
16	2 39.894	2 49.724	2 59.553	3 09.383	3 19.212	3 29.042	3 38.871	3 48.701	16 .044
17	2 40.058	2 49.888	2 59.717	3 09.547	3 19.376	3 29.206	3 39.035	3 48.865	17 .046
18	2 40.222	2 50.051	2 59.881	3 09.710	3 19.540	3 29.370	3 39.199	3 49.029	18 .049
19	2 40.386	2 50.215	3 00.045	3 09.874	3 19.704	3 29.533	3 39.363	3 49.193	19 .052
20	2 40.549	2 50.379	3 00.209	3 10.038	3 19.868	3 29.697	3 39.527	3 49.356	20 .055
21	2 40.713	2 50.543	3 00.372	3 10.202	3 20.032	3 29.861	3 39.691	3 49.520	21 .057
22	2 40.877	2 50.707	3 00.536	3 10.366	3 20.195	3 30.025	3 39.854	3 49.684	22 .060
23	2 41.041	2 50.870	3 00.700	3 10.530	3 20.359	3 30.189	3 40.018	3 49.848	23 .063
24	2 41.205	2 51.034	3 00.864	3 10.693	3 20.523	3 30.353	3 40.182	3 50.012	24 .066
25	2 41.369	2 51.198	3 01.028	3 10.857	3 20.687	3 30.516	3 40.346	3 50.175	25 .068
26	2 41.532	2 51.362	3 01.192	3 11.021	3 20.851	3 30.680	3 40.510	3 50.339	26 .071
27	2 41.696	2 51.526	3 01.355	3 11.185	3 21.014	3 30.844	3 40.674	3 50.503	27 .074
28	2 41.960	2 51.690	3 01.519	3 11.349	3 21.178	3 31.008	3 40.837	3 50.667	28 .076
29	2 42.024	2 51.853	3 01.683	3 11.513	3 21.342	3 31.172	3 41.001	3 50.831	29 .079
30	2 42.188	2 52.017	3 01.847	3 11.676	3 21.506	3 31.336	3 41.165	3 50.995	30 .082
31	2 42.352	2 52.181	3 02.011	3 11.840	3 21.670	3 31.499	3 41.329	3 51.158	31 .085
32	2 42.515	2 52.345	3 02.174	3 12.004	3 21.834	3 31.663	3 41.493	3 51.322	32 .087
33	2 42.679	2 52.509	3 02.338	3 12.168	3 21.997	3 31.827	3 41.657	3 51.486	33 .090
34	2 42.843	2 52.673	3 02.502	3 12.332	3 22.161	3 31.991	3 41.820	3 51.650	34 .093
35	2 43.007	2 52.836	3 02.666	3 12.496	3 22.325	3 32.155	3 41.984	3 51.814	35 .096
36	2 43.171	2 53.000	3 02.830	3 12.659	3 22.489	3 32.318	3 42.148	3 51.978	36 .098
37	2 43.334	2 53.164	3 02.994	3 12.823	3 22.653	3 32.482	3 42.312	3 52.141	37 .101
38	2 43.498	2 53.328	3 03.157	3 12.987	3 22.817	3 32.646	3 42.476	3 52.305	38 .104
39	2 43.662	2 53.492	3 03.321	3 13.151	3 22.980	3 32.810	3 42.639	3 52.469	39 .105
40	2 43.826	2 53.656	3 03.485	3 13.315	3 23.144	3 32.974	3 42.803	3 52.633	40 .109
41	2 43.990	2 53.819	3 03.649	3 13.478	3 23.308	3 33.138	3 42.967	3 52.797	41 .112
42	2 44.154	2 53.983	3 03.813	3 13.642	3 23.472	3 33.301	3 43.131	3 52.961	42 .115
43	2 44.317	2 54.147	3 03.977	3 13.806	3 23.636	3 33.465	3 43.295	3 53.124	43 .117
44	2 44.481	2 54.311	3 04.140	3 13.970	3 23.800	3 33.629	3 43.459	3 53.288	44 .120
45	2 44.645	2 54.475	3 04.304	3 14.134	3 23.963	3 33.793	3 43.622	3 53.452	45 .123
46	2 44.809	2 54.638	3 04.468	3 14.298	3 24.127	3 33.957	3 43.786	3 53.616	46 .126
47	2 44.973	2 54.802	3 04.632	3 14.461	3 24.291	3 34.121	3 43.950	3 53.780	47 .128
48	2 45.137	2 54.966	3 04.796	3 14.625	3 24.455	3 34.284	3 44.114	3 53.943	48 .131
49	2 45.300	2 55.130	3 04.960	3 14.789	3 24.619	3 34.448	3 44.278	3 54.107	49 .134
50	2 45.464	2 55.294	3 05.123	3 14.953	3 24.782	3 34.612	3 44.442	3 54.271	50 .137
51	2 45.628	2 55.458	3 05.287	3 15.117	3 24.946	3 34.776	3 44.605	3 54.435	51 .139
52	2 45.792	2 55.621	3 05.451	3 15.281	3 25.110	3 34.940	3 44.769	3 54.599	52 .142
53	2 45.956	2 55.785	3 05.615	3 15.444	3 25.274	3 35.104	3 44.933	3 54.763	53 .145
54	2 46.120	2 55.949	3 05.779	3 15.608	3 25.438	3 35.267	3 45.097	3 54.926	54 .147
55	2 46.283	2 56.113	3 05.942	3 15.772	3 25.602	3 35.431	3 45.261	3 55.090	55 .150
56	2 46.447	2 56.277	3 06.106	3 15.936	3 25.765	3 35.595	3 45.425	3 55.254	56 .153
57	2 46.611	2 56.441	3 06.270	3 16.100	3 25.929	3 35.759	3 45.588	3 55.418	57 .156
58	2 46.775	2 56.604	3 06.434	3 16.264	3 26.093	3 35.923	3 45.752	3 55.582	58 .158
59	2 46.939	2 56.768	3 06.598	3 16.427	3 26.257	3 36.086	3 45.916	3 55.746	59 .161

Barre Duparcq's Military Art and History. Elements of Military Art and History: comprising the History and Tactics of the separate Arms; the Combination of the Arms; and the minor operations of War. By EDWARD DE LA BARRE DUPARCQ, Chef de Bataillon of Engineers in the Army of France; and Professor of the Military Art in the Imperial School of St. Cyr. Translated by Brig.-Gen. GEO. W. CULLUM, U. S. A., Chief of the Staff of Major-Gen. H. W. HALLECK, General-in-Chief U. S. Army. 1 vol. 8vo, cloth. $5.00.

' I read the original a few years since, and considered it the very best work I had seen upon the subject. Gen. Cullum's ability, and familiarity with the technical language of French military writers, are a sufficient guarantee of the correctness of his translation." H. W. HALLECK, Major-Gen. U. S. A.

Benet's Military Law. A Treatise on Military Law and the Practice of Courts-Martial. By Capt. S. V. BENÉT, Ordnance Department, U. S. A., late Assistant Professor of Ethics, Law, &c., Military Academy, West Point. Fifth edition, revised. 1 vol. 8vo, law sheep. $4.50.

" This book is manifestly well timed just at this particular period, and it is, without doubt, quite as happily adapted to the purpose for which it was written. It is arranged with admirable method, and written with such perspicuity, and in a style so easy and graceful, as to engage the attention of every reader who may be so fortunate as to open its pages. This treatise will make a valuable addition to the library of the lawyer or the civilian; while to the military man it seems to be indispensable."—*Philadelphia Evening Journal.*

Halleck's International Law. International Law; or, Rules Regulating the Intercourse of States in Peace and War. By Major-Gen. H. W. HALLECK, Commanding the Army. 1 vol. 8vo, law sheep. $6.00.

" The work will be found to be of great use to army and navy officers, to professional lawyers, and to all interested in the topics of which it treats—topics to which present events give a greatly enhanced importance—such as 'Declaration of War and its Effects;' 'Sieges and Blockades;' 'Visitation and Search;' 'Right of Search;' 'Prize Courts;' 'Military Occupation;' 'Treaties of Peace;' 'Sovereignty of States,' &c., and valuable information for consuls and ambassadors."—*N. Y. Evening Post.*

History of West Point. And its Military Importance during the American Revolution; and the Origin and Progress of the United States Military Academy. By Capt. EDWARD C. BOYNTON, A. M., Adjutant of the Military Academy. With numerous Maps and Engravings. 1 vol. 8vo, blue cloth, $6.00; half mor., $7.50; full mor., $10.00.

" It records the earliest attempt at instituting a Military School by the Continental Congress, in 1776. It conducts us through the life of the institution, arguing with terseness its constitutionality, defending its educational principles, and explaining the necessity for its preservation."—*United Service Magazine.*

1

History of the United States Naval Academy. With Biographical Sketches, and the Names of all the Superintendents, Professors, and Graduates; to which is added a Record of some of the earliest votes by Congress, of Thanks, Medals, and Swords to Naval Officers. By EDWARD CHAUNCEY MARSHALL, A. M. 1 vol. 12mo, cloth, plates. $1.00.

"Every naval man will find it not only a pleasant companion, but an invaluable book of reference. It is seldom that so much information is made accessible in so agreeable a manner in so small a space."—*New York Times.*

Scott's Military Dictionary. Comprising Technical Definitions; Information on Raising and Keeping Troops; Actual Service, including Makeshifts and improved *Matériel,* and Law, Government, Regulation, and Administration relating to Land Forces. By Colonel H. L. Scott, Inspector-General U. S. A. 1 vol. large octavo, fully illustrated, half morocco, $6; half russia, $8; full morocco, $10.

"This book is really an Encyclopædia, both elementary and technical, and as such occupies a gap in military literature which has long been most inconveniently vacant. This book meets a present popular want, and will be secured not only by those embarking in the profession but by a great number of civilians, who are determined to follow the descriptions and to understand the philosophy of the various movements of the Campaign. Indeed, no tolerably good library would be complete without the work."—*New York Times.*

Benton's Ordnance and Gunnery. A Course of Instruction in Ordnance and Gunnery; compiled for the use of the Cadets of the United States Military Academy. By Capt. J. G. Benton, Ordnance Department, late Instructor of Ordnance and Gunnery, Military Academy, West Point; Principal Assistant to Chief of Ordnance, U. S. A. Third Edition, revised and enlarged. 1 vol. 8vo, cloth, cuts. $5.00.

"There is no one book within the range of our military reading and study, that contains more to recommend it upon the subject of which it treats. It is as full and complete as the narrow compass of a single volume would admit, and the reputation of the author as a scientific and practical artillerist is a sufficient guarantee for the correctness of his statements and deductions, and the thoroughness of his labors."—*N. Y. Observer.*

Gibbon's Artillerist's Manual. Compiled from various Sources, and adapted to the Service of the United States. Profusely illustrated with woodcuts and engravings on stone. Second edition, revised and corrected, with valuable additions. By Gen. John Gibbon, U. S. A. 1 vol. 8vo, half roan. $6.00.

This book is now considered the standard authority for that particular branch of the Service in the United States Army. The War Department, at Washington, has exhibited its thorough appreciation of the merits of this volume, the want of which has been hitherto much felt in the service, by subscribing for 700 copies.

Jomini's Life of the Emperor Napoleon I. Life of Napoleon. By Baron Jomini, General-in-Chief and Aide-de-Camp to the Emperor of Russia. Translated from the French, with Notes, by H. W. Halleck, LL. D., Major-General United States Army; author of "Elements of Military Art and Science," "International Law, and the Laws of War,'' etc., etc. In four volumes octavo, with an Atlas of Sixty Maps and Plans. Price, in red cloth, $25.00; half calf, or half morocco, $35.00; half russia, $37.50.

"It is needless to say any thing in praise of Jomini as a writer on the science of war.

"General Halleck has laid the professional soldier and the student of military history under equal obligations by the service he has done to the cause of military literature in the preparation of this work for the press. His rare qualifications for the task thus undertaken will be acknowledged by all.

"The Notes with which the text is illustrated by General Halleck are not among the least of the merits of the publication, which, in this respect, has a value not possessed by the original work. * * *" —*National Intelligencer.*

"The Atlas attached to this version of Jomini's *Napoleon* adds very materially to its value. It contains *sixty* Maps, illustrative of Napoleon's extraordinary military career, beginning with the immortal Italian Campaigns of 1796, and closing with the decisive Campaign of Flanders, in 1815, the last Map showing the Battle of Wavre. These Maps take the reader to Italy, Egypt, Palestine, Germany Moravia, Russia, Spain, Portugal, and Flanders; and their number and variety, and the vast and various theatres of action which they indicate, testify to the immense extent of Napoleon's operations, and to the gigantic character of his power. * * *" —*Boston Traveller.*

Jomini's Grand Military Operations. Treatise on Grand Military Operations; or, a Critical and Military History of the Wars of Frederick the Great, as contrasted with the Modern System, together with a few of the most important Principles of the Art of War. By Baron Jomini, General-in-Chief and Aide-de-Camp to the Emperor of Russia. Translated from the French by Col. S. B. Holabird, A. D. C. U. S. A. 2 vols. 8vo, cloth, with an Atlas of 40 Maps and Plans. $15.00.

Jomini's Campaign of Waterloo. The Political and Military History of the Campaign of Waterloo. Translated from the French of General Baron de Jomini, by Capt. S. V. Benét, Ordnance Department, U. S. Army. Third edition. 1 vol. 12mo, cloth, $1.25.

"Baron Jomini has the reputation of being one of the greatest military historians and critics of the century. His merits have been recognized by the highest military authorities in Europe, and were rewarded in a conspicuous manner by the greatest military power in Christendom. He learned the art of war in the school of experience, the best and only finishing school of the soldier. He served with distinction in nearly all the campaigns of Napoleon, and it was mainly from the gigantic military operations of this matchless master of the art that he was enabled to discover its true principles, and to ascertain the best means of their application to the infinity of combinations which actual war presents. Jomini criticises the details of Waterloo with great science, and yet in a manner that interests the general reader as well as the professional." —*New York World.*

Roemer's Cavalry; its History, Management, and Uses in War. By J. Roemer, LL. D., late an Officer of Cavalry in the Service of the Netherlands. Elegantly illustrated, with one hundred and twenty-seven fine wood engravings. In one large octavo volume, beautifully printed on tinted paper. Cloth, $6; half calf, $7.50.

"By far the best treatise upon Cavalry and its uses in the field, which has yet been published in this country, for the general use of officers of all ranks, is this elaborate and interesting work. Eschewing the elementary principles and tactics of cavalry, which may be learned from any hand-book, the author treats of the uses of cavalry in the field of strategy and tactics, and of its general discipline and management. The range of the work includes an admirable treatise upon rifled fire-arms, an historical sketch of cavalry, embodying many interesting facts, an account of the cavalry service in Europe and this country, and a treatise on horses, their equipment, management, &c. The work is copiously illustrated and elegantly printed. It is interesting not alone to military men but to the general reader, who will gain from its pages valuable historical facts and very clear ideas of some branches of the art of war, such as the employment of spies, gaining information in an enemy's country, advance movements, and other strategical manœuvres."—*Boston Journal.*

Nolan's System for Training Cavalry Horses. By Kenner Garrard, Captain Fifth Cavalry, U. S. A. 1 vol. 12mo, cloth. 24 lithographed plates. $2.00.

"This work is clearly written, is eminently practical, is fully illustrated, and contains numerous hints as applicable to the discipline and management of the draught-horse as that of his more showy and fiery brother of the cavalry."—*Boston Journal.*

Barnard and Barry's Peninsular Campaign. Report of the Engineer and Artillery Operations of the Army of the Potomac, from its organization to the close of the Peninsular Campaign. By Brig.-Gen. J. G. Barnard, and other Engineer Officers, and Brig.-Gen. W. F. Barry, Chief of Artillery. Illustrated by numerous Maps, Plans, etc. 1 vol. 8vo, cloth, $4.00.

"The title of this work sufficiently indicates its importance and value as a contribution to the history of the great rebellion. General Barnard's Report is a narrative of the Engineer operations of the Army of the Potomac from the time of its organization to the date it was withdrawn from the James River. Thus a record is given of an important part in the great work which the nation found before it when it was first confronted with the necessity of war; and perhaps on no other point in the annals of the rebellion will future generations look with a deeper or more admiring interest."—*Buffalo Courier.*

The "C. S. A.," and the Battle of Bull Run. (A Letter to an English friend), by J. G. Barnard, Lieut.-Colonel of Engineers, U. S. A., Brigadier-General and Chief Engineer, Army of the Potomac. With five Maps. 1 vol. 8vo, cloth. $2.00.

"The work is clearly written, and can but leave the impression upon every reader's mind that it is truth. We commend it to the perusal of every one who wants an intelligent, truthful, and graphic description of the 'C. S. A.,' and the Battle of Bull Run."—*New York Observer.*

D. Van Nostrand's Publications.

Simpson's Ordnance and Naval Gunnery. A Treatise on Ordnance and Naval Gunnery, compiled and arranged as a Text-Book for the U. S. Naval Academy. By Lieut. Edward Simpson, U. S. N. Third edition, revised and enlarged. 1 vol. 8vo, plates and cuts, cloth. $5.00.

"It is scarcely necessary for us to say, that a work prepared by a writer so practically conversant with all the subjects of which he treats, and who has such a reputation for scientific ability, cannot fail to take at once a high place among the text-books of our naval service. It has been approved by the Secretary of the Navy, and will henceforth be one of the standard authorities on all matters connected with Naval Gunnery."—*New York Herald.*

Holley's Ordnance and Armor. Embracing Descriptions, Discussions, and Professional Opinions concerning the Material, Fabrication, Requirements, Capabilities, and Endurance of European and American Guns for Naval, Sea-Coast, and Iron-Clad Warfare, and their Rifling, Projectiles, and Breech-Loading; also, Results of Experiments against Armor, from Official Records. With an Appendix, referring to Gun-Cotton, Hooped Guns, etc., etc. By A. L. Holley, B. P. With 493 illustrations. 1 vol. 8vo, 948 pages. Half roan, $10.00.

Luce's Naval Light Artillery. Instructions for Naval Light Artillery, afloat and ashore, prepared and arranged for the U. S. Naval Academy, by Lieutenant W. H. Parker, U. S. N. Second edition, revised by Lieutenant S. B. Luce, U. S. N., Assistant Instructor of Gunnery and Tactics at the U. S. Naval Academy. 1 vol. 8vo, cloth, with twenty-two Plates. $3.00.

"The service for which this is the text-book of instruction is of special importance in the present war. The use of light boat-pieces is constant and important, and young officers are frequently obliged to leave their boats, take their pieces ashore, and manœuvre them as field artillery. Not unfrequently, also, they are incorporated, when ashore, with troops, and must handle their guns like the artillery soldiers of a battery. 'The Exercise of the Howitzer Afloat' was prepared and arranged by Captain Dahlgren, whose name gives additional sanction and value to the book. A Manual for the Sword and Pistol is also given.' The Plates are numerous and exceedingly clear, and the whole typography is excellent."—*Philadelphia Inquirer.*

Ward's Naval Ordnance and Gunnery. Elementary Instruction in Naval Ordnance and Gunnery. By James H. Ward, Commander U. S. Navy; author of "Naval Tactics," and "Steam for the Million." New edition, revised and enlarged. 8vo, cloth. $2.00.

"It conveys an amount of information in the same space to be found nowhere else, and given with a clearness which renders it useful as well to the general as the professional inquirer."—*N. Y. Evening Post.*

"The whole detail of Ordnance, in its history, philosophy, and application, is given by Commander Ward in such a manner (with occasional diagrams) as to convey to the student accurate notions for practical use."—*New Yorker.*

5

Luce's Seamanship. Compiled from various authorities, and Illustrated with numerous original and selected Designs. For the use of the United States Naval Academy. By S. B. Luce, Lieut.-Commander U. S. N. In two parts. Second Edition. One royal octavo volume, cloth. $10.00.

Squadron Tactics Under Steam. By Foxhall A. Parker, Commander U. S. Navy. Published by authority of the Navy Department. 1 vol. 8vo, with numerous Plates. $5.00.

"In this useful work to Navy officers, the author demonstrates—by the aid of profuse diagrams and explanatory text—a new principle for manœuvring naval vessels in action. The author contends that the winds, waves, and currents of the ocean oppose no more serious obstacles to the movements of a steam fleet, than do the inequalities on the surface of the earth to the manœuvres of an army. It is in this light, therefore, that he views a vast fleet—simply as an army; the regiments, brigades, and divisions of which are represented by a certain ship or ships."—*Scientific American.*

Nautical Routine and Stowage. With Short Rules in Navigation. By John McLeod Murphy and Wm. N. Jeffers, Jr., U. S. Navy. 1 vol. 8vo, blue cloth. $2.50.

Osbon's Hand-Book of the United States Navy. Being a Compilation of all of the Principal Events in the History of every Vessel of the United States Navy, from April, 1861, to May, 1864. Compiled and arranged by B. S. Osbon. 1 vol. 12mo, blue cloth. $2.50.

"As a condensed and compact history, as well as a work containing a vast amount of information, this work cannot be surpassed."—*Boston Traveller.*

Brandt's Gunnery Catechism. Gunnery Catechism, as applied to the Service of Naval Ordnance. Adapted to the Latest Official Regulations, and approved by the Bureau of Ordnance, Navy Department. By J. D. Brandt, formerly of the U. S. Navy. 1 vol. 18mo, illustrated, blue cloth. $1.50.

"This manual is very full of information and instruction, and shows the 'chief end' of Gunnery, and the aim of those who follow that profession. It is indispensable to all those who are suddenly introduced to a gun-deck, and will be found a valuable aid also to experienced officers."—*Commercial Advertiser.*

Barrett's Gunnery Instructions. Gunnery Instructions, simplified for the Volunteer Officers of the United States Navy, with Hints to Executive and other Officers. By Lieut. Edward Barrett, U. S. N., Instructor of Gunnery, Navy Yard, Brooklyn. 1 vol. 12mo, cloth. $1.25.

"It is a thorough work, treating plainly on its subject, and contains also some valuable hints to executive officers. No officer in the volunteer navy should be without a copy."—*Boston Evening Traveller.*

D. Van Nostrand's Publications.

Totten's Naval Text-Book. Naval Text-Book and Dictionary, compiled for the use of the Midshipmen of the U. S. Navy. By Com mander B. J. Totten, U. S. N. Second and revised edition. 1 vol. 12mo. $3.00.

Calculated Tables of Ranges for Navy and Army Guns. With a Method of Finding the Distance of an Object at Sea. By Lieut. W. P Buckner, U. S N. 1 vol. 8vo, cloth. $1.50.

Manual of Internal Rules and Regulations for Men-of-War. By Commodore U. P. Levy, U. S. N., late Flag Officer commanding U. S. Naval Force in the Mediterranean, &c. Flexible blue cloth. Third Edition, revised and enlarged. 50 cents.

"Among the professional publications for which we are indebted to the war, we willingly give a prominent place to this useful little Manual of Rules and Regulations to be observed on board of ships of war. Its authorship is a sufficient guarantee for its accuracy and practical value; and as a guide to young officers in providing for the discipline, police, and sanitary government of the vessels under their command, we know of nothing superior."—*N. Y. Herald.*

King's Lessons and Practical Notes on Steam, The Steam Engine, Propellers, &c., &c., for Young Marine Engineers, Students, and others. By the late W. H. King, U. S. Navy. Revised by Chief Engineer J. W. King, U. S. Navy. Ninth Edition, enlarged. 8vo, cloth. $2.00.

"This is the ninth edition of a valuable work of the late W. H. King, U. S. Navy. It contains lessons and practical notes on Steam and the Steam-Engine, Propellers, &c. It is calculated to be of great use to young marine engineers, students, and others. The text is illustrated and explained by numerous diagrams and representations of machinery. This new edition has been revised and enlarged by Chief Engineer J. W. King, U. S. Navy, brother to the deceased author of the work."—*Boston Daily Advertiser.*

Ward's Steam for the Million. A popular Treatise on Steam and its Application to the Useful Arts, especially to Navigation. By J. H. Ward, Commander U. S Navy. New and revised Edition. 1 vol. 8vo, cloth. $1.00.

The Naval Howitzer Ashore. By Foxhall A. Parker, Commander U. S. Navy. 1 vol. 8vo, with Plates. Cloth. $4.00.

7

D. Van Nostrand's Publications.

Gillmore's Fort Sumter. Official Report of Operations against the Defences of Charleston Harbor, 1863. Comprising the Descent upon Morris Island, the Demolition of Fort Sumter, and the Siege and Reduction of Forts Wagner and Gregg. By Major-Gen. Q. A. GILLMORE, U. S. Volunteers, and Major U. S. Corps of Engineers. With Maps and Lithographic Plates, Views, &c. 1 vol. 8vo. cloth. $10.

Gillmore's Siege and Reduction of Fort Pulaski, Georgia. Papers on Practical Engineering. No. 8. Official Report to the U. S. Engineer Department of the Siege and Reduction of Fort Pulaski, Ga., February, March, and April, 1862. By Brig.-Gen. Q. A. GILLMORE, U. S. A. Illustrated by Maps and Views. 1 vol. 8vo, cloth. $2.50.

"This is an official history of the siege of Fort Pulaski, from the commencement, with all the details in full, made up from a daily record, forming a most valuable paper for future reference. The situation and construction of the Fort, the position of the guns, both of the rebels and the Federals, and their operation, are made plain by maps and engraved views of different sections. Additional reports from other officers are furnished in the appendix, and every thing has been done to render the work full and reliable."—*Boston Journal.*

Gillmore's Treatise on Limes, Hydraulic Cements, and Mortars. Papers on Practical Engineering, U. S. Engineer Department, No. 9, containing Reports of numerous Experiments conducted in New York City, during the years 1858 to 1861, inclusive. By Q. A. GILLMORE, Brig.-Gen. U. S. Volunteers, and Major U. S. Corps of Engineers. With numerous Illustrations. One vol. 8vo. $4.00.

"This work contains a record of certain experiments and researches made under the authority of the Engineer Bureau of the War Department from 1858 to 1861, upon the various hydraulic cements of the United States, and the materials for their manufacture. The experiments were carefully made, and are well reported and compiled."—*Journal Franklin Institute.*

The Volunteer Quartermaster. Containing a Collection and Codification of the Laws, Regulations, Rules, and Practices governing the Quartermaster's Department of the United States Army, and in force March 4, 1865. By Captain ROELIFF BRINKERHOFF, Assistant Quartermaster U. S. Volunteers, and Post Quartermaster at Washington. 1 vol. 12mo, cloth. $2.50.

This work embraces all the laws of Congress, and all the orders and circulars of the War Office and its bureaus, bearing upon the subject. It also embodies the decisions of the Second Comptroller of the Treasury, so far as they affect the Quartermaster's Department. These decisions have the force of law in the adjustment of accounts, and are therefore invaluable to all disbursing officers.

8

Cullum's Military Bridges. Systems of Military Bridges in Use by the United States Army; those adopted by the Great European Powers; and such as are employed in British India. With Directions for the Preservation, Destruction, and Re-establishment of Bridges. By Brig.-Gen. GEORGE W. CULLUM, Lieut.-Col. Corps of Engineers, U. S. A. 1 vol. 8vo. With numerous Illustrations. cloth. $3.50.

"We have no man more competent to prepare such a work than Brig.-Gen. Cullum, who had the almost exclusive supervision, devising, building, and preparing for service of the various bridge-trains sent to our armies in Mexico during our war with that country. The treatise before us is very complete, and has evidently been prepared with scrupulous care. The descriptions of the various systems of military bridges adopted by nearly all civilized nations are very interesting even to the non-professional reader, and to those specially interested in such subjects must be very instructive, for they are evidently the work of a master of the art of military bridge-building."— *Washington Chronicle.*

Haupt's Military Bridges. For the Passage of Infantry, Artillery, and Baggage-Trains; with Suggestions of many new Expedients and Constructions for Crossing Streams and Chasms; designed to utilize the Resources ordinarily at command, and reduce the amount and cost of Army Transportation. Including also Designs for Trestle and Truss Bridges for Military Railroads, adapted especially to the wants of the Service of the United States. By HERMAN HAUPT, Brig.-Gen. in charge of the Construction and Operation of the U. S. Military Railways, Author of "General Theory of Bridge Construction," &c. Illustrated by sixty-nine Lithographic Engravings. 8vo, cloth. $6.50.

"This elaborate and carefully prepared, though thoroughly practical and simple work, is peculiarly adapted to the military service of the United States. Mr. Haupt has added very much to the ordinary facilities for crossing streams and chasms, by the instructions afforded in this work."— *Boston Courier.*

Holley's Railway Practice. American and European Railway Practice, in the Economical Generation of Steam, including the Materials and Construction of Coal-burning Boilers, Combustion, the Variable Blast, Vaporization, Circulation, Superheating, Supplying and Heating Feed-water, &c., and the Adaptation of Wood and Coke-burning Engines to Coal-burning; and in permanent Way, including Road-bed, Sleepers, Rails, Joint Fastenings, Street Railways, &c., &c. By Alexander L. Holley, B. P. With 77 lithographed plates. 1 vol. folio, cloth. $12.00.

* * * "All these subjects are treated by the author in both an intelligent and intelligible manner. The facts and ideas are well arranged, and presented in a clear and simple style, accompanied by beautiful engravings, and we presume the work will be regarded as indispensable by all who are interested in a knowledge of the construction of railroads, and rolling stock, or the working of locomotives."—*Scientific American.*

Authorized U. S. Infantry Tactics. For the Instruction, Exercise, and Manœuvres of the Soldier, a Company, Line of Skirmishers, Battalion, Brigade, or Corps d'Armée. By Brig.-Gen. SILAS CASEY, U S. A. 3 vols 24mo. Cloth, lithographed plates. $2.50.

Vol. I.—School of the Soldier; School of the Company; Instruction for Skirmishers.

Vol. II.—School of the Battalion.

Vol. III.—Evolutions of a Brigade; Evolutions of a Corps d'Armée.

"WAR DEPARTMENT, WASHINGTON, *August* 11, 1862.

"The System of Infantry Tactics prepared by Brig.-Gen. Silas Casey, U. S. A., having been approved by the President, is adopted for the Instruction of the Infantry of the Armies of the United States, whether Regular, Volunteer, or Militia.

"EDWIN M. STANTON, *Secretary of War.*"

U. S. Tactics for Colored Troops. U. S. Infantry Tactics, for the Instruction, Exercise, and Manœuvres of the Soldier, a Company, Line of Skirmishers, and Battalion, for the use of the Colored Troops of the United States Infantry. Prepared under the direction of the War Department. 1 vol., plates. $1.50.

"WAR DEPARTMENT, WASHINGTON, *March* 9, 1863.

"This system of United States Infantry Tactics, prepared under the direction of the War Department, for the use of the Colored Troops of the United States Infantry, having been approved by the President, is adopted for the instruction of such troops.

"EDWIN M. STANTON, *Secretary of War.*"

Kelton's Bayonet Exercise. A New Manual of the Bayonet, for the Army and Militia of the United States. By Colonel J. C. KELTON, U. S. A. With forty beautifully engraved plates. Red cloth. $2.00.

"This Manual was prepared for the use of the Corps of Cadets, and has been introduced at the Military Academy with satisfactory results. It is simply the theory of the attack and defence of the sword applied to the bayonet, on the authority of men skilled in the use of arms."—*New York Times.*

Berriman's Sword-Play. The Militiaman's Manual and Sword-Play without a Master.—Rapier and Broad-Sword Exercises copiously Explained and Illustrated; Small-Arm Light Infantry Drill of the United States Army; Infantry Manual of Percussion Muskets; Company Drill of the United States Cavalry. By Major M. W. BERRIMAN, engaged for the last thirty years in the practical instruction of Military Students. Fourth edition. 1 vol. 12mo, red cloth. $1.00.

"This work will be found very valuable to all persons seeking military instruction; but it recommends itself most especially to officers, and those who have to use the sword or sabre. We believe it is the only work on the use of the sword published in this country."—*New York Tablet.*

Heavy Artillery Tactics.—1863. Instruction for Heavy Artillery; prepared by a Board of Officers, for the use of the Army of the
United States. With service of a gun mounted on an iron carriage. In
one volume, 12mo, with numerous illustrations. cloth. $2.50.

> "WAR DEPARTMENT,
> "WASHINGTON, D. C., *Oct* 20, 1862.

"This system of Heavy Artillery Tactics, prepared under direction of the War Department, having been approved by the President, is adopted for the instruction of
troops when acting as heavy artillery. EDWIN M. STANTON, *Secretary of War.*"

"The First Part consists of sixteen lessons relating to the service of the single piece, including the gun, howitzer, mortar, and columbiad; also, the formation of batteries, the art
of aiming pieces and firing hot-shot. Part Second relates entirely to mechanical manœuvres, and appliances for handling, mounting, dismounting, and transporting heavy pieces.
Part Third is of a miscellaneous character, containing directions for embarking and disem-
barking artillery and ordnance stores; also, tables of dimensions and weights of guns,
carriages, shot, shell, machines, and implements, with charges for, and ranges of heavy
artillery. These instructions are not only copious in detail, but aptly illustrated with
thirty-nine elegant steel-plate engravings."—*Bulletin.*

Roberts's Hand-Book of Artillery. For the Service of the
United States Army and Militia. New revised and greatly enlarged
edition. By Major JOSEPH ROBERTS, U. S. A. 1 vol. 18mo, cloth.
$1.25.

"A complete catechism of gun practice, covering the whole ground of this branch of
military science, and adapted to militia and volunteer drill, as well as to the regular
army. It has the merit of precise detail, even to the technical names of all parts of a
gun, and how the smallest operations connected with its use can be best performed. It
has evidently been prepared with great care, and with strict scientific accuracy."
New York Century.

Duane's Manual for Engineer Troops. Consisting of Part
I., Pontoon Drill; II., Practical Operations of a Siege; III, School of
the Sap; IV., Military Mining; V., Construction of Batteries. By
Major J. C. DUANE, Corps of Engineers, U. S. Army. 1 vol. 12mo,
cloth. $2.50.

"I have carefully examined Capt. J. C. Duane's 'Manual for Engineer Troops,' and do
not hesitate to pronounce it the very best work on the subject of which it treats.
"H. W. HALLECK, *Major-General, U. S. A.*"

Dufour's Principles of Strategy and Grand Tactics.
Translated from the French of General G. H. Dufour. By WILLIAM P.
CRAIGHILL, Capt. of Engineers, U. S. Army, and Assistant Professor of
Engineering, U. S. Military Academy, West Point. From the last
French Edition. Illustrated. In one volume 12mo. cloth. $3.00.

"In all military matters General Dufour is recognized as one of the first authorities in
Europe, and consequently the translation of this very valuable work is a most acceptable
addition to our military libraries."—*London Naval and Military Gazette.*

Instructions for Field Artillery. Prepared by a Board of Artillery Officers. 1 vol. 12mo, illustrated by 122 pages of Engravings. Cloth. $3.00.

"WAR DEPARTMENT, WASHINGTON, *March* 1, 1863.

"This system of Instruction for Field Artillery, prepared under direction of the War Department, having been approved by the President, is adopted for the instruction of troops when acting as field artillery.

"Accordingly, instruction in the same will be given after the method pointed out therein; and all additions to or departures from the exercise and manœuvres laid down in the system, are positively forbidden.

"EDWIN M. STANTON, *Secretary of War.*"

Anderson's Evolutions of Field Batteries of Artillery. Translated from the French, and arranged for the Army and Militia of the United States. By Gen. ROBERT ANDERSON, U. S. A. Published by order of the War Department. 1 vol., cloth, 32 plates. $1.

"WAR DEPARTMENT, *Nov.* 2d, 1859.

"The system of 'Evolutions of Field Batteries,' translated from the French, and arranged for the service of the United States, by Major Robert Anderson, of the 1st Regiment of Artillery, having been approved by the President, is published for the information and government of the army.

"All Evolutions of Field Batteries not embraced in this system are prohibited, and those herein prescribed will be strictly observed. J. B. FLOYD,

"*Secretary of War.*"

Mendell's Treatise on Military Surveying. Theoretical and Practical, including a Description of Surveying Instruments. By G. H. MENDELL, Captain of Engineers. 1 vol. 8vo, with numerous illustrations. cloth. $2.50.

"The author is a Captain of Engineers, and has for his chief authorities Salneuve, Lalobre, and Simms. He has presented the subject in a simple form, and has liberally illustrated it with diagrams, that it may be readily comprehended by every one who is liable to be called upon to furnish a military sketch of a portion of country."—*New York Evening Post.*

Viele's Hand-book. Hand-Book for Active Service, containing Practical Instructions in Campaign Duties. For the use of Volunteers. By Brigadier-General Egbert L. Viele, U. S. A. 12mo, cloth. $1.

"It is a thorough treatise, copiously illustrated, and embraces a complete drill by company, regiment, &c. It also embraces instructions in regard to the camp, fortifications, rations, and mode of cooking them, and has a manual for light and heavy artillery."—*New Haven Palladium.*

A Treatise on the Camp and March. With which is connected the Construction of Field-Works and Military Bridges; with an Appendix of Artillery Ranges, &c. For the use of Volunteers and Militia in the United States. By Captain Henry D. Grafton, U. S. A. 1 vol. 12mo, cloth. 75 cents.

The Automaton Regiment; or, Infantry Soldier's Practical Instructor.—For all Regimental Movements in the Field. By G. Douglas Brewerton, U. S. Army. Neatly put up in boxes, price $1; when sent by mail, $1.40.

The "Automaton Regiment" is a simple combination of blocks and counters, so arranged and designated, by a carefully considered contrast of colors, that it supplies the student with a perfect miniature regiment, in which the position in the battalion of each company, and of every officer and man in each division, company, platoon, and section, is clearly indicated. It supplies the studious soldier with the means whereby he can consult his "tactics," and at the same time join practice to theory by manœuvring a mimic regiment.

I hereby certify that I have examined the "Automaton Regiment," invented by G. Douglas Brewerton, late of the United States Regular Army, and now serving as a Volunteer Aide upon my military staff, and believe that his invention will prove a useful and valuable assistant to every student of military tactics. I take pleasure in recommending it accordingly.   B. SAXTON,

Brigadier-General Volunteers.

The Automaton Company; or, Infantry Soldier's Practical Instructor.—For all Company Movements in the Field. By G. Douglas Brewerton, U. S. A. Price in boxes, $1.25; when sent by mail, $1.95.

The Automaton Battery; or, Artillerist's Practical Instructor.—For all Mounted Artillery Manœuvres in the Field. By G. Douglas Brewerton, U. S. A. Price in boxes, $1; when sent by mail, $1.40.

These productions are of a similar character, and cannot fail to be of great value to the military student. They are *object lessons*, that will teach more in an hour than mere verbal instruction could in a month. They cannot be too highly commended to both officers and men.—*Commercial Advertiser.*

Monroe's Company and Skirmish Drill.—The Company Drill of the Infantry of the Line, together with the Skirmish Drill of the Company and Battalion, after the method of General Le Loutorel. Bayonet Fencing; with a Supplement on the Handling and Service of Light Infantry. By J. Monroe, Colonel 22d Regiment, N. G., N. Y. S. M., formerly Captain U. S. Infantry. 1 vol. 32mo, cloth. 75 cents.

This is a most valuable and timely little manual. It should be in the hands of every new recruit.—*Chicago Tribune.*

School of the Guides.—Designed for the use of the Militia of the United States. By Colonel Eugene Le Gal, 55th Regiment, N. Y. S. M. cloth, 60 cents.

"This excellent compilation condenses into a compass of less than sixty pages all the instruction necessary for the guides, and the information being disconnected with other matters, is more readily referred to and more easily acquired."—*Louisville Journal.*

13

Craighill's Army Officer's Pocket Companion. Principally designed for Staff Officers in the Field. Partly translated from the French of M. DE ROUVRE, Lieutenant-Colonel of the French Staff Corps, with Additions from standard American, French, and English Authorities. By WM. P. CRAIGHILL, First Lieutenant U. S. Corps of Engineers, Assist. Prof. of Engineering at the U. S. Military Academy, West Point. 1 vol. 18mo, full roan. $2.00.

"I have carefully examined Capt. Craighill's Pocket Companion. I find it one of the very best works of the kind I have ever seen. Any Army or Volunteer officer who will make himself acquainted with the contents of this little book, will seldom be ignorant of his duties in camp or field.   H. W. HALLECK, *Major-General U. S. A.*"

Hunter's Manual for Quartermasters and Commissaries. Containing Instructions in the Preparation of Vouchers, Abstracts, Returns, etc., embracing all the recent changes in the Army Regulations, together with instructions respecting Taxation of Salaries, etc. By Captain R. F. HUNTER, late of the U. S. Army. 12mo, cloth, $1.25. Flexible morocco, $1.50.

"This is the only work of the kind extant. It is based on the latest regulations of the War Department, and will be regarded as authority by those officers for whose use it is designed."—*Saturday Evening Gazette.*

Ordronaux's Manual of Instructions for Military Surgeons, in the Examination of Recruits and Discharge of Soldiers. With an Appendix, containing the Official Regulations of the Provost-Marshal General's Bureau, and those for the formation of the Invalid Corps, etc., etc. Prepared at the request of the United States Sanitary Commission. By JOHN ORDRONAUX, M. D., Professor of Medical Jurisprudence in Columbia College, New York. 12mo. Half morocco. $1.50.

"In a condensed form, it is an admirable treatise on the important subjects of which it treats. The author has aimed to be brief without being obscure, to omit nothing of real importance, and to draw his materials from the best sources. He treats of the physical disabilities which have relation to the military service, and of these alone. Medical Examiners are instructed in their duties, and the method of discovering feigned, artificially produced, and concealed diseases is pointed out. The book will prove valuable to all who are concerned in the manipulation of recruits or conscripts. An Appendix contains official regulations and instructions relative to the Provost-Marshal's office, the Invalid Corps, etc."—*Commercial Advertiser.*

Ordronaux's Hints on the Preservation of Health in Armies. For the use of Volunteer Officers and Soldiers. By JOHN ORDRONAUX, M. D. New edition, 18mo, cloth. 75 cents.

14

Thomas's Rifled Ordnance. A Practical Treatise on the Application of the Principle of the Rifle to Guns and Mortars of every Calibre. To which is added, a new theory of the initial action and force of Fired Gunpowder. By LYNALL THOMAS, F. R. S. L. Fifth edition, revised. One volume, octavo, illustrated. · cloth· $2.00.

"At a time when the manufacture of guns engrosses the attention of thousands on thousands, any practical treatise which may suggest desirable alterations or innovations is of importance, and deserves that attention we doubt not will be extended to the present volume."—*Boston Evening Gazette.*

Wilcox's Rifles and Rifle Practice. An Elementary Treatise on the Theory of Rifle Firing; explaining the causes of Inaccuracy of Fire and the manner of correcting it; with descriptions of the Infantry Rifles of Europe and the United States, their Balls and Cartridges. By Captain C. M. WILCOX, U. S. A. New edition, with engravings and cuts. Green cloth. $2.00.

"The book will be found intensely interesting to all who are watching the changes in the art of war arising from the introduction of the new rifled arms. We recommend to our readers to buy the book."—*Military Gazette.*

Lendy's Maxims and Instructions on the Art of War. Maxims, Advice, and Instructions on the Art of War; or, A Practical Military Guide for the use of Soldiers of all Arms and of all Countries. Translated from the French by Captain LENDY, Director of the Practical Military College, late of the French Staff, etc., etc. 1 vol. 18mo, cloth. 75 cents.

"This book treats generally of the art of war and the conduct of campaigns, and without going into all the details of a soldier's business, aims to explain the principles on which an army may be well established in camp, or successfully led and manœuvred on the field."—*Providence Journal.*

Andrews's Hints to Company Officers on their Military Duties. By Captain C. C. ANDREWS, Third Regiment Minnesota Volunteers. 1 vol. 18mo, cloth. 60 cents.

"This is a hand-book of good practical advice, which officers of all ranks may study with advantage."—*Philadelphia Press.*

Heth's System of Target Practice.—For the use of Troops when armed with the Musket, Rifle-Musket, Rifle, or Carbine. Prepared principally from the French, by Captain Henry Heth, 10th Infantry, U S. A. 18mo, cloth. 75 cents.

15

D. Van Nostrand's Publications.

Minifie's Mechanical Drawing. A Text-Book of Geometrical Drawing, for the Use of Mechanics and Schools, in which the Definitions and Rules of Geometry are familiarly explained; the Practical Problems are arranged from the most simple to the more complex, and in their description technicalities are avoided as much as possible. With Illustrations for Drawing Plans, Sections, and Elevations of Buildings and Machinery; an introduction to Isometrical Drawing, and an Essay on Linear Perspective and Shadows. Illustrated with over 200 Diagrams engraved on Steel. By WILLIAM MINIFIE, Architect. Seventh edition. With an Appendix on the Theory and Application of Colors. 1 vol. 8vo, cloth. $4.00.

"It is the best work on Drawing that we have ever seen, and is especially a text-book of Geometrical Drawing for the use of Mechanics and Schools. No young Mechanic, such as a Machinist, Engineer, Cabinet-Maker, Millwright, or Carpenter, should be without it.' —*Scientific American.*

"One of the most comprehensive works of the kind ever published, and cannot but possess great value to builders. The style is at once elegant and substantial."—*Pennsylvania Inquirer.*

"We think this the *best* work on this subject, which is saying very much ; as much attention has been given to the science of Drawing. There is nothing in the range of drawing that cannot be found in this book, and which is not well explained."—*Ohio Teacher.*

"Whatever is said is rendered perfectly intelligible by remarkably well executed diagrams on steel, leaving nothing for mere vague supposition; and the addition of an introduction to isometrical drawing, linear perspective, and the projection of shadows, winding up with a useful index to technical terms."—*Glasgow Mechanic's Journal.*

Minifie's Geometrical Drawing. Abridged from the Octavo edition, for the use of Schools. Illustrated with 48 Steel Plates. Fifth edition. 1 vol. 12mo, half roan. $1.50.

Peirce's System of Analytic Mechanics. Physical and Celestial Mechanics, by BENJAMIN PEIRCE, Perkins Professor of Astronomy and Mathematics in Harvard University, and Consulting Astronomer of the American Ephemeris and Nautical Almanac. Developed in four systems of Analytic Mechanics, Celestial Mechanics, Potential Physics, and Analytic Morphology. 1 vol 4to, cloth. $10.00.

Dictionary of Weights and Measures. Universal Dictionary of Weights and Measures, Ancient and Modern, reduced to the standards of the United States of America. By J. H. ALEXANDER. New edition, enlarged. 1 vol. 8vo, cloth. $2.50.

16

www.ingramcontent.com/pod-product-compliance
Lightning Source LLC
Chambersburg PA
CBHW032153010726
47493CB00008BA/2677